# Vegetable Dishes

Second Supplement to the Fifth Edition of

*McCance and Widdowson's*

# The Composition of Foods

## COPYRIGHT

# Vegetable Dishes

Second Supplement to the Fifth Edition of

McCance and Widdowson's

## The Composition of Foods

B. Holland, A.A. Welch and D.H. Buss

The Royal Society of Chemistry
and
Ministry of Agriculture, Fisheries and Food

The Royal Society of Chemistry
Thomas Graham House
Science Park
Milton Road
Cambridge CB4 4WF
UK

Tel.:   (0223) 420066     Telex: 818293
Fax:   (0223) 423429


ISBN 0-85186-396-5




*Orders should be addressed to:*
The Royal Society of Chemistry
Turpin Distribution Services Ltd
Letchworth
Herts. SG6 1HN

Tel.:   (0462) 67255     Telex: 825372
Fax:   (0462) 480947




Photocomposed by Land and Unwin, Bugbrooke


Printed in the United Kingdom by
Richard Clay Ltd, Bungay, Suffolk

# CONTENTS

# ACKNOWLEDGEMENTS

A large number of people have helped at each stage in the preparation of this book.

We are indebted to numerous manufacturers, retailers and other organisations for recipe details and additional information on the range and composition of their products. In particular we would like to thank the Aga Khan Health Board for the United Kingdom, Brooke Bond Foods Ltd, Crestway Marketing Ltd, Dalepak Foods plc, Food Enterprises Ltd, Mrs Gill's Indian Kitchen, the Haldane Foods Group, Marlow Foods Ltd, Raj Foods, St Ivel Ltd, Safeway plc, Sahib Foods Ltd, J Sainsbury plc, J A Sharwood and Co Ltd, Tesco Stores Ltd, Tivall Frozen Foods and Waitrose Ltd.

We are also very grateful to many dietitians and nutritionists for providing recipe details from both personal and survey collections as well as for their useful suggestions and comments. Thanks go to Shelley Ahmed from Westminster Social Services, Jane Brophy from The Vegetarian Society of the United Kingdom Ltd, Eleanor Carlson from Lifeline Nutritional Services, Jasmine Challis from The Middlesex Hospital, dietitians from Harrow, Leicestershire, Newham, Southern Derbyshire, and Wandsworth Health Authorities, Anne Earless from Leeds Polytechnic, Elaine Gardner from Waltham Forest College, Georgina Holt from the University of Bradford, Pat Judd and Tashmin Kassam from King's College London, Mary Morgan from the Vegetarian Catering Advisory Service and researchers at King's College Hospital, London.

In addition we would like to thank Aarti Datani, freelance nutritionist; Smita Ganatra, paediatric dietitian, Central Middlesex Hospital; Azmina Govindji, chief dietitian, British Diabetic Association; and Aruna Thacker, senior dietitian, St George's Hospital Tooting for their valuable contributions to the contents of this supplement as well as their time spent checking the final recipes.

The new analyses of vegetable dishes and products included in this supplement were performed at the Laboratory of the Government Chemist, Teddington under the direction of Mrs G D Holcombe and at Leeds University under the direction of Mrs J Ryley. The non-starch polysaccharide fractions were determined at the Dunn Clinical Nutrition Centre by a team headed by Dr H N Englyst.

The National Dairy Council and Vegetarian Society kindly provided a number of the cover photographs.

The final preparation of this book was overseen by a committee which, besides the authors, comprised Miss P J Brereton (Northwick Park Hospital, Harrow), Dr M C Edwards (Campden Food and Drink Research Association, Chipping Campden), Dr A M Fehily (MRC Epidemiology Unit, Cardiff), Miss A A Paul (MRC Dunn Nutritional Laboratory, Cambridge), Professor D A T Southgate (AFRC Institute of Food Research, Norwich) and Dr F J Taylor (Royal Society of Chemistry, Cambridge).

We would also like to express our appreciation for all the help given to us by so many people in the Ministry of Agriculture, Fisheries and Food, the Royal Society of Chemistry and elsewhere who were involved in the work leading up to the production of this book. This supplement is the first to be produced from a new production system set up at the Royal Society of Chemistry for which we thank Julie Moncur, David Stout and Stewart White from Computer Services Division.

Special thanks are extended to Mr I D Unwin, Food Information Consultant, for his valuable contributions to the data from which this supplement has been produced as well as his earlier work on the nutrient databank project and associated publications.

# INTRODUCTION

This book presents information on the amounts of more than 50 different nutrients in a wide variety of vegetable dishes eaten in the UK. This complements the nutritional data on the 401 raw, processed, and cooked vegetables included in the supplement on *Vegetables, Herbs and Spices* (Holland *et al.*, 1991a), for no information on the nutrients in vegetable dishes was included in that book or in the fifth edition of *The Composition of Foods* (Holland *et al.*, 1991b).

The number and variety of vegetable dishes eaten in the UK has increased greatly in recent years with the growth in vegetarianism and in the number of people from the Indian sub-continent. There is also an increasing number of commercially-available vegetable-based dishes. This book thus contains information on 347 such dishes, including some related nut dishes, and cereal-based dishes containing vegetables as an important ingredient. It should be noted that a number of the dishes contain animal products such as milk or eggs, but none contains any meat or fish.

This book has been produced by the Ministry of Agriculture, Fisheries and Food and the Royal Society of Chemistry, who are updating and extending the information on the nutritional value of foods in the UK. The results of this work are published in three formats. The most comprehensive consists of a series of detailed books, of which this is the fifth, each presenting values for a wide range of nutrients in a large number of foods in a particular food group. These are supplements to the more wide ranging textbook on *The Composition of Foods*, the fifth edition of which was published in 1991 (Holland *et al.*, 1991b) and contains information on the major nutrients in a wide selection of foods of all types. All the information from the supplements, including this one, and from the fifth edition is also available in computer-readable format, details of which can be obtained from the Royal Society of Chemistry.

## Sources of the data
### Recipes

There are few literature values on the nutrients in vegetable dishes, so most of the values in this supplement have been derived by calculation from recipes using information on the composition of the ingredients and any changes of weight in cooking due to losses of water by evaporation or gains of water or fat by absorption. Vitamin losses on cooking have also been taken into account.

Full details of each recipe or the proportions of the constituents in the dishes are given in an extensive Appendix (page 135). Every attempt has been made to ensure that the recipes are as representative as possible, but users are reminded of the importance of allowing for any differences between these and the dishes they are evaluating. The recipes were obtained from a wide variety of sources including food manufacturers, retailers and caterers; from many dietitians, research workers and others who have conducted quantitative dietary surveys on representative groups of the British population; and from standard recipe books. Where more than one recipe is commonly used, the nutritional value has been calculated for each of the main variations such as after preparation with different kinds of flour, milk or fat. The recipes were then checked or prepared by two different people, who measured any gains or losses of water or gains in fat.

Most of the nutrient values for the component vegetables and pulses used in these dishes were taken from the supplement on *Vegetables, Herbs and Spices*, while those for the other ingredients were taken from the fifth edition and associated supplements. If, however, more recent values were available, these were used instead. Vitamin losses on cooking were then calculated by a computer using the appropriate values from Tables 16, 17 and 21 in the fifth edition of *The Composition of Foods*, with the losses in vegetables having been derived during recent analytical studies at the Norwich Laboratory of the Institute of Food Research. More details of the principles used in calculating the nutrients in recipe dishes are given in section 3.4 of the fifth edition.

### By analysis

The proportions of the ingredients in some of the manufactured products, ready-to-eat foods and other dishes were unknown and so their composition could not be determined by calculation. In such cases, arrangements were made for representative samples of the food to be analysed at the Laboratory of the Government Chemist or the Procter Department of Food Science, Leeds University. Care was taken to ensure that the samples were fully representative of the products described. The analytical methods were as described in the fifth edition of The Composition of Foods.

## Arrangement of the tables

Because of the wide range of ingredients and therefore of nutrients in many vegetable dishes, the information for each food is presented in six pages instead of the four pages used in previous supplements.

### Food names and grouping

The dishes have been listed in alphabetical order under their most common name. Bhajis and curries (which differ mainly in that the latter contain more added water) are each grouped together, but for these and for all other recipes the sequence has as far as possible been determined by the name of the main or characterising ingredient. However, many vegetable dishes have more than one such ingredient and many have more than one common name, so an Appendix (page 208) gives alternative names for many of the dishes. There is also a list of alternative names for many of the ingredients and a detailed index to help with identification and coding. The systematic Latin names of the vegetables and nuts can be found in the supplement on *Vegetables, Herbs and Spices* and the supplement on *Fruit and Nuts* (Holland *et al.*, 1992) respectively.

### Numbering system

The foods have been numbered in sequence from 1 to 347, and, as in previous supplements, each also has a unique two digit prefix. The prefix for the foods in this supplement is '15', so the full code number for, for example, aubergine stuffed with lentils and vegetables (the first food in this book) is 15-001. This is the number that will be used in nutrient databank applications.

*Description and main data sources*

The information given under this heading includes a more detailed description of many of the dishes. The main source of information on the composition of each is also given, and where different recipes are used by different ethnic groups, the appropriate ethnic group is indicated. The source and number of samples taken for chemical analysis has also been included in this section.

*Nutrients*

All nutrient values are given per 100 grams of the dish as ready to eat. Although peel and other inedible material had to be removed from many of the ingredients prior to preparation, there is no obligatory wastage from any of the final dishes. The edible proportion of all the dishes is therefore 1.00, and all the values are the same as the amounts of nutrients per 100g edible portion.

*Proximates*: – The second page for each food begins with the water content of the dish, and this is followed by the amounts of nitrogen, protein, fat, available carbohydrate expressed as its monosaccharide equivalent, and energy value both in kilocalories and kilojoules. Protein was derived from the nitrogen values by multiplying them by 6.25, and the energy values were derived by multiplying the amounts of protein, fat and carbohydrate by the factors in **Table 1**.

**Table 1**  *Energy conversion factors*

|  | kcal/g | kJ/g |
|---|---|---|
| Protein | 4 | 17 |
| Fat | 9 | 37 |
| Available carbohydrate<br>  expressed as monosaccharide | 3.75 | 16 |

These values for the energy derived from carbohydrate are different from, and scientifically more accurate than, those required for the nutrition labelling of foods. Reference should be made to *Food Labelling Data for Manufacturers* (RSC/MAFF, 1992) for further details of values to be used for this purpose.

*Carbohydrates*: – The third page gives more details of the individual carbohydrates. The value for total sugars is the sum of the glucose, fructose, galactose, sucrose, maltose and lactose in the food, but does not include raffinose, stachyose or other oligosaccharides that are present in significant quantities in many pulses and other vegetables. The total amount of these oligosaccharides is given separately, but these have been included together with the sugars and starch in the total amount of carbohydrate shown on the preceding page. As in the previous UK tables, the amounts of sugars, starch and total carbohydrate have all been shown after conversion to their monosaccharide equivalents.

*Fibre and fat fractions*: – The fourth page for each food first shows the main dietary fibre fractions and the amount of phytic acid present in the dish. These fibre values have not been converted to their monosaccharide equivalents but are the actual amounts of each component present. The relationships between the various forms and fractions of fibre are shown in **Table 2**.

**Table 2**  *Relationships between the dietary fibre fractions*

Cellulose

Insoluble non-cellulosic polysaccharides ⎫ Insoluble fibre

Soluble non-cellulosic polysaccharides ⎫ Soluble fibre

'Lignin'

⎫ Englyst fibre (non-starch polysaccharides)

⎫ Southgate fibre[a] (unavailable carbohydrate)

[a] The Southgate values are generally higher than NSP values because they include substances measuring as lignin and also because the enzymatic preparation used leaves some enzymatically resistant starch in the dietary fibre residue. A 'resistant starch' value can be obtained from the NSP procedures but because this uses different conditions and enzymes this may or may not be the same as the enzymatically resistant starch in the Southgate method.

Since most vegetable recipes contain fat, the fourth page also shows the amounts of saturated, monounsaturated and polyunsaturated fatty acids and the cholesterol content of each food. Cholesterol values are given in milligrams; to convert to mmol, divide by 386.6. Where the dish can commonly be prepared with different oils or fats, values for each are given. The margarine used contained only vegetable oils, and the blended cooking oil consisted mainly of rapeseed and soya oils.

*Minerals and vitamins*: – The range of minerals and vitamins shown on the fifth and sixth pages for each food is the same as in the supplement on *Vegetables, Herbs and Spices*, except for the omission of sulfur. Sodium and chloride will vary according to the amount of salt used; in general, the vegetables in these recipes were assumed to have been boiled in unsalted water. The values for total carotene and for vitamin E take account of the relative activities of the different fractions of each that are known to be present in the ingredients. The amounts of these fractions in the main ingredients are given in the supplements on *Vegetables, Herbs and Spices* and on *Fruit and Nuts*.

## Appendices

Full details of all the recipes used in this supplement are given on page 135. There is also a list of the more common alternative names for many of the dishes on page 208, to complement the information given in the index. The third Appendix gives the full code numbers and alternative names for the ingredients used in the recipes, so that the exact sources of the nutrient data used in the recipe calculations can be identified.

## Nutrient variability

Any vegetable or other ingredient used in these dishes can vary markedly in composition. Some nutrients will differ in a consistent way between varieties of a vegetable and with season, but there may be greater differences with length of storage and, for certain vegetables, with the depth of peeling or number of outer leaves removed. But differences arising from the method of cultivation (such as 'organic' methods) appear to be small and inconsistent.

The nutrient content of the dishes included in this supplement can also be markedly affected by variations in the proportions of the ingredients used. Although care has been taken to ensure that each recipe is as representative as possible of the dish described, different people may use different amounts of the ingredients, or may substitute one ingredient for another, or may even omit one or more ingredients altogether. The amount of water and salt used, and the length of cooking, also affect the nutritional value, although there will be little or no difference between dishes cooked with microwaves or by more conventional methods.

It is not practical to give different nutrient values for most of these factors, so the tables in general show average values for each dish. It is therefore important that these tables are not used uncritically, and that the user is aware of any major differences between the dishes they are evaluating and the recipes described here.

## References to Introductory text

Holland, B., Unwin, I.D. and Buss, D.H (1988) Third supplement to McCance and Widdowson's The Composition of Foods, 4th edition: Cereals and Cereal Products, Royal Society of Chemistry, Cambridge

Holland, B., Unwin, I.D. and Buss, D.H (1989) Fourth supplement to McCance and Widdowson's The Composition of Foods, 4th edition: Milk Products and Eggs, Royal Society of Chemistry, Cambridge

Holland, B., Unwin, I.D. and Buss, D.H. (1991a) Fifth supplement to McCance and Widdowson's The Composition of Foods, 4th edition: Vegetables, Herbs and Spices, Royal Society of Chemistry, Cambridge

Holland, B., Unwin, I.D. and Buss, D.H. (1992) First supplement to McCance and Widdowson's The Composition of Foods, 5th edition: Fruit and Nuts, Royal Society of Chemistry, Cambridge

Holland, B., Welch, A.A., Unwin, I.D., Buss, D.H., Paul, A.A and Southgate, D.A.T. (1991b) McCance and Widdowson's The Composition of Foods 5th edition, Royal Society of Chemistry, Cambridge

RSC/MAFF (1992) Food Labelling Data for Manufacturers, Royal Society of Chemistry, Cambridge

# The
# Tables

# Symbols and abbreviations used in the tables

## Symbols

| | |
|---|---|
| 0 | None of the nutrient is present |
| Tr | Trace |
| N | The nutrient is present in significant quantities but there is no reliable information on the amount |
| ( ) | Estimated value |

## Abbreviations

| | |
|---|---|
| Trypt | Tryptophan |
| Satd | Saturated |
| Monounsatd | Monounsaturated |
| Polyunsatd | Polyunsaturated |

| No. 15- | Food | Description and main data sources |
|---------|------|-----------------------------------|
| 1 | **Aubergine**, stuffed with lentils and vegetables | Recipe from dietary survey records. Whole dish consumed |
| 2 | stuffed with rice | Rice, tomato, onion and raisin stuffing. Recipe from review of recipe collection. Whole dish consumed |
| 3 | stuffed with vegetables, cheese topping | Tomato, mushroom and onion stuffing. Recipe from review of recipe collection. Whole dish consumed |
| 4 | **Bean loaf** | Recipe from dietary survey records. Mixed beans |
| 5 | **Beanburger**, aduki, *fried in vegetable oil* | Recipe adapted from recipe No 8 |
| 6 | butter bean, *fried in vegetable oil* | Recipe from review of recipe collection |
| 7 | red kidney bean, *fried in vegetable oil* | Recipe from Leeds Polytechnic |
| 8 | soya, *fried in vegetable oil* | Recipe from review of recipe collection |
| 9 | **Bhaji**, aubergine and potato | Recipe from review of recipe collection |
| 10 | aubergine, pea, potato and cauliflower | Gujerati dish. Recipe from a personal collection |
| 11 | cabbage | Recipe adapted from Carlson and de Wet, 1991 |
| 12 | cabbage and pea, *with butter ghee* | Bangladeshi dish. Recipe adapted from Carlson and de Wet, 1991 |
| 13 | *with vegetable oil* | Bangladeshi dish. Recipe adapted from Carlson and de Wet, 1991 |

# Vegetable Dishes

**Composition of food per 100g**

| No. 15- | Food | Water g | Total nitrogen g | Protein g | Fat g | Carbohydrate g | Energy value kcal | kJ |
|---|---|---|---|---|---|---|---|---|
| 1 | **Aubergine**, stuffed with lentils and vegetables | 85.5 | 0.34 | 2.0 | 4.6 | 3.9 | 64 | 267 |
| 2 | stuffed with rice | 77.0 | 0.29 | 1.8 | 2.4 | 14.8 | 84 | 355 |
| 3 | stuffed with vegetables, cheese topping | 74.4 | 0.91 | 5.3 | 9.8 | 4.3 | 124 | 518 |
| 4 | **Bean loaf** | 54.5 | 1.16 | 6.9 | 7.3 | 15.1 | 149 | 627 |
| 5 | **Beanburger**, aduki, *fried in vegetable oil* | 54.6 | 1.33 | 8.2 | 7.2 | 22.9 | 183 | 772 |
| 6 | butter bean, *fried in vegetable oil* | 58.8 | 0.95 | 5.7 | 10.8 | 19.2 | 191 | 802 |
| 7 | red kidney bean, *fried in vegetable oil* | 55.4 | 1.03 | 6.3 | 10.8 | 21.7 | 203 | 853 |
| 8 | soya, *fried in vegetable oil* | 57.2 | 1.74 | 10.7 | 11.0 | 13.7 | 193 | 806 |
| 9 | **Bhaji**, aubergine and potato | 73.4 | 0.31 | 2.0 | 8.8 | 12.0 | 130 | 540 |
| 10 | aubergine, pea, potato and cauliflower | 78.6 | 0.53 | 3.3 | 2.6 | 9.5 | 70 | 298 |
| 11 | cabbage | 80.7 | 0.38 | 2.3 | 4.9 | 6.7 | 77 | 320 |
| 12 | cabbage and pea, *with butter ghee* | 67.2 | 0.53 | 3.3 | 14.7 | 9.2 | 178 | 739 |
| 13 | *with vegetable oil* | 67.2 | 0.53 | 3.3 | 14.7 | 9.2 | 178 | 739 |

**Vegetable Dishes**

**Carbohydrate fractions, g per 100g food**

| No. 15- | Food | Starch | Total sugars | Individual sugars | | | | | | Oligo-saccharides |
|---|---|---|---|---|---|---|---|---|---|---|
| | | | | Glucose | Fructose | Galactose | Sucrose | Maltose | Lactose | |
| 1 | **Aubergine**, stuffed with lentils and vegetables | 2.1 | 1.7 | 0.8 | 0.7 | 0 | 0.1 | 0 | 0 | 0.1 |
| 2 | stuffed with rice | 8.8 | 5.9 | 3.0 | 2.7 | 0 | 0.2 | 0 | 0 | 0.2 |
| 3 | stuffed with vegetables, cheese topping | 0.2 | 3.8 | 1.8 | 1.6 | 0 | 0.4 | 0 | Tr | 0.3 |
| 4 | **Bean loaf** | 10.0 | 3.9 | 1.2 | 1.1 | 0 | 1.6 | 0 | 0 | 1.2 |
| 5 | **Beanburger**, aduki, *fried in vegetable oil* | 19.2 | 2.6 | 0.8 | 0.7 | 0 | 1.0 | 0.1 | 0 | 1.1 |
| 6 | butter bean, *fried in vegetable oil* | 15.2 | 3.1 | 0.8 | 0.6 | 0 | 1.6 | 0 | 0 | 1.0 |
| 7 | red kidney bean, *fried in vegetable oil* | 16.2 | 4.4 | 0.8 | 0.7 | 0 | 2.7 | Tr | 0 | 1.1 |
| 8 | soya, *fried in vegetable oil* | 9.1 | 3.5 | 0.9 | 0.8 | 0 | 1.8 | 0 | 0 | 1.0 |
| 9 | **Bhaji**, aubergine and potato | 8.4 | 2.9 | 1.2 | 0.9 | 0 | 0.8 | 0 | 0 | 0.7 |
| 10 | aubergine, pea, potato and cauliflower | 5.4 | 3.6 | 0.9 | 0.7 | 0 | 2.0 | 0 | 0 | 0.6 |
| 11 | cabbage | 0.5 | 5.5 | 2.5 | 2.2 | 0 | 0.9 | 0 | 0 | 0.7 |
| 12 | cabbage and pea, *with butter ghee* | 1.6 | 5.9 | 2.2 | 1.9 | 0 | 1.8 | 0 | Tr | 1.8 |
| 13 | *with vegetable oil* | 1.6 | 5.9 | 2.2 | 1.9 | 0 | 1.8 | 0 | 0 | 1.8 |

# Vegetable Dishes

**Fibre fractions, phytic acid and fatty acids, g per 100g food**
**Cholesterol, mg per 100g food**

| No. 15- | Food | Dietary fibre, g | | Fibre fractions, g | | | | Phytic acid g | Fatty acids, g | | | Cholesterol mg |
| | | Southgate method | Englyst method | Cellulose | Non-cellulosic polysaccharide Soluble | Insoluble | Lignin | | Satd | Mono-unsatd | Poly-unsatd | |
|---|---|---|---|---|---|---|---|---|---|---|---|---|
| 1 | **Aubergine**, stuffed with lentils and vegetables | (2.2) | 1.9 | 0.6 | 0.8 | 0.5 | N | 0.03 | 0.5 | 1.6 | 2.2 | 0 |
| 2 | stuffed with rice | 2.9 | 2.1 | 0.7 | 1.0 | 0.3 | 0.3 | 0.05 | 0.4 | 0.8 | 1.1 | 0 |
| 3 | stuffed with vegetables, cheese topping | 2.9 | 2.3 | 0.8 | 1.0 | 0.5 | 0.2 | 0.03 | 3.7 | 3.0 | 2.5 | 14 |
| 4 | **Bean loaf** | (4.6) | 4.3 | 1.2 | 1.6 | 1.5 | N | (0.24) | 0.8 | 2.2 | 3.5 | 0 |
| 5 | **Beanburger**, aduki, *fried in vegetable oil* | 4.4 | 4.3 | 1.2 | 1.2 | 1.9 | N | N | 1.0 | 2.6 | 3.0 | 32 |
| 6 | butter bean, *fried in vegetable oil* | (4.1) | 4.1 | 1.0 | 2.0 | 1.1 | N | 0.29 | 1.2 | 3.7 | 5.1 | 0 |
| 7 | red kidney bean, *fried in vegetable oil* | (6.1) | 4.9 | 1.1 | 2.5 | 1.3 | N | 0.29 | 1.2 | 3.7 | 5.1 | 0 |
| 8 | soya, *fried in vegetable oil* | (4.7) | 4.7 | 1.2 | 2.2 | 1.3 | N | 0.37 | 1.5 | 3.3 | 4.9 | 32 |
| 9 | **Bhaji**, aubergine and potato | 2.0 | 1.7 | 0.6 | 0.9 | 0.2 | 0.1 | 0.01 | 0.9 | 3.1 | 4.1 | 0 |
| 10 | aubergine, pea, potato and cauliflower | 3.6 | 2.9 | 1.2 | 1.2 | 0.5 | 0.1 | 0.02 | 0.3 | 0.8 | 1.2 | 0 |
| 11 | cabbage | (3.3) | 2.7 | 0.9 | 1.4 | 0.4 | 0.4 | 0.02 | 0.6 | 1.6 | 2.2 | 0 |
| 12 | cabbage and pea, *with butter ghee* | (4.3) | (3.4) | (1.5) | (1.5) | (0.5) | 0.3 | 0.02 | 9.4 | 3.5 | 0.7 | 39 |
| 13 | *with vegetable oil* | (4.3) | (3.4) | (1.5) | (1.5) | (0.5) | 0.3 | 0.02 | 1.6 | 5.1 | 7.0 | 0 |

# Vegetable Dishes

## Inorganic constituents per 100g food

| No. 15- | Food | Na | K | Ca | Mg | P | Fe | Cu | Zn | Cl | Mn | Se | I |
|---|---|---|---|---|---|---|---|---|---|---|---|---|---|
| | | | | | | mg | | | | | | µg | |
| 1 | **Aubergine**, stuffed with lentils and vegetables | 79 | 240 | 12 | 12 | 38 | 0.8 | 0.11 | 0.3 | 31 | 0.2 | (7) | N |
| 2 | stuffed with rice | 170 | 290 | 20 | 16 | 38 | 0.6 | 0.07 | 0.4 | 270 | 0.2 | (2) | 3 |
| 3 | stuffed with vegetables, cheese topping | 250 | 390 | 120 | 20 | 110 | 0.7 | 0.20 | 0.7 | 430 | 0.2 | (5) | 9 |
| 4 | **Bean loaf** | 140 | 410 | 40 | 37 | 130 | 1.8 | 0.27 | 0.8 | 310 | 0.5 | 7 | 2 |
| 5 | **Beanburger**, aduki, *fried in vegetable oil* | 220 | 480 | 46 | 52 | 180 | 2.1 | 0.38 | 1.9 | 52 | 1.0 | 2 | N |
| 6 | butter bean, *fried in vegetable oil* | 300 | 310 | 31 | 35 | 130 | 1.7 | 0.16 | 0.9 | 470 | 0.8 | N | N |
| 7 | red kidney bean, *fried in vegetable oil* | 290 | 310 | 61 | 36 | 160 | 2.0 | 0.09 | 1.0 | 460 | 0.8 | 4 | N |
| 8 | soya, *fried in vegetable oil* | 220 | 450 | 69 | 54 | 220 | 2.7 | 0.28 | 1.1 | 54 | 0.9 | 4 | N |
| 9 | **Bhaji**, aubergine and potato | 280 | 330 | 20 | 19 | 38 | 0.8 | 0.08 | 0.3 | 470 | 0.2 | (1) | 4 |
| 10 | aubergine, pea, potato and cauliflower | 330 | 350 | 23 | 24 | 58 | 1.2 | (0.05) | 0.6 | 530 | 0.3 | (1) | 2 |
| 11 | cabbage | 220 | 340 | 59 | 12 | 54 | 0.9 | 0.04 | 0.4 | 370 | 0.2 | (1) | 4 |
| 12 | cabbage and pea, *with butter ghee* | 250 | 320 | 53 | 19 | 66 | 1.2 | (0.05) | 0.6 | 410 | 0.3 | (1) | 3 |
| 13 | *with vegetable oil* | 250 | 320 | 53 | 19 | 66 | 1.2 | (0.05) | 0.6 | 410 | 0.3 | (1) | 5 |

# Vegetable Dishes

| No. 15- | Food | Retinol µg | Carotene µg | Vitamin D µg | Vitamin E mg | Thiamin mg | Ribo-flavin mg | Niacin mg | Trypt 60 mg | Vitamin B6 mg | Vitamin B12 µg | Folate µg | Panto-thenate mg | Biotin µg | Vitamin C mg |
|---|---|---|---|---|---|---|---|---|---|---|---|---|---|---|---|
| 1 | **Aubergine**, stuffed with lentils and vegetables | 0 | 125 | 0 | 1.21 | 0.04 | 0.03 | 0.5 | 0.3 | 0.09 | 0 | 10 | N | N | 3 |
| 2 | stuffed with rice | 0 | 180 | 0 | 0.62 | 0.04 | 0.01 | 0.5 | 0.3 | 0.11 | 0 | 10 | (0.12) | N | 4 |
| 3 | stuffed with vegetables, cheese topping | 46 | 345 | Tr | 1.81 | 0.07 | 0.11 | 1.1 | 1.1 | 0.16 | 0.2 | 18 | 0.54 | N | 5 |
| 4 | **Bean loaf** | 0 | 170 | 0 | 2.21 | 0.14 | 0.07 | 1.3 | 1.2 | 0.18 | 0 | 30 | 0.46 | 7.5 | 3 |
| 5 | **Beanburger**, aduki, *fried in vegetable oil* | 15 | 160 | 0.2 | N | 0.19 | 0.11 | 1.1 | 1.6 | N | 0.2 | N | N | N | 5 |
| 6 | butter bean, *fried in vegetable oil* | 0 | 1435 | 0 | 2.72 | 0.20 | 0.22 | 1.4 | 1.1 | 0.12 | 0 | 16 | N | N | 1 |
| 7 | red kidney bean, *fried in vegetable oil* | 0 | 1440 | 0 | 2.65 | 0.27 | 0.19 | 1.3 | 1.2 | 0.15 | 0 | 17 | 0.25 | N | 1 |
| 8 | soya, *fried in vegetable oil* | 15 | 160 | 0.2 | 2.48 | 0.18 | 0.11 | 0.9 | 2.0 | 0.18 | 0.2 | 22 | 0.48 | 18.1 | 5 |
| 9 | **Bhaji**, aubergine and potato | 0 | 97 | 0 | 2.26 | 0.13 | 0.02 | 0.5 | 0.4 | 0.25 | 0 | 14 | 0.21 | N | 5 |
| 10 | aubergine, pea, potato and cauliflower | 0 | 250 | 0 | 0.56 | 0.15 | 0.04 | 0.7 | 0.7 | 0.19 | 0 | 26 | 0.25 | N | 10 |
| 11 | cabbage | 0 | 545 | 0 | 1.33 | 0.15 | 0.02 | 0.6 | 0.4 | 0.19 | 0 | 37 | 0.18 | 0.3 | 24 |
| 12 | cabbage and pea, *with butter ghee* | 94 | (475) | 0.3 | (0.78) | 0.19 | 0.03 | 1.0 | 0.6 | 0.18 | Tr | 35 | 0.18 | 0.5 | 17 |
| 13 | *with vegetable oil* | 0 | (405) | 0 | (3.70) | 0.19 | 0.03 | 1.0 | 0.6 | 0.18 | 0 | 35 | 0.18 | 0.5 | 17 |

| No.<br>15- | Food | Description and main data sources |
|---|---|---|
| 14 | **Bhaji,** cabbage and potato, *with butter* | Punjabi dish. Recipe from dietary survey records |
| 15 | *with vegetable oil* | Punjabi dish. Recipe from dietary survey records |
| 16 | cabbage and spinach | Punjabi dish. Cabbage, spinach, onion and chick pea flour. Recipe from manufacturer |
| 17 | carrot, potato and pea, *with butter* | Punjabi dish. Recipe from dietary survey records |
| 18 | *with vegetable oil* | Punjabi dish. Recipe from dietary survey records |
| 19 | cauliflower | Recipe from Wharton et al, 1983 |
| 20 | cauliflower and potato | Bangladeshi dish. Recipe adapted from Carlson and de Wet, 1991 |
| 21 | cauliflower, potato and pea, *with butter* | Punjabi dish. Recipe from manufacturer |
| 22 | *with vegetable oil* | Punjabi dish. Recipe from manufacturer |
| 23 | cauliflower and vegetable | Cauliflower, onion, tomato and spices. Recipe from Wharton et al, 1983 |
| 24 | green bean | Punjabi dish. Recipe from manufacturer |
| 25 | karela, *with butter ghee* | Recipe from a personal collection |
| 26 | *with vegetable oil* | Recipe from a personal collection |
| 27 | mushroom | Mushroom and onion. Recipe from review of recipe collection |
| 28 | mustard leaves | Recipe from Wharton et al, 1983 |
| 29 | mustard leaves and spinach | Recipe from Wharton et al, 1983 |

# Vegetable Dishes

Composition of food per 100g

| No. 15- | Food | Water g | Total nitrogen g | Protein g | Fat g | Carbohydrate g | Energy value kcal | kJ |
|---|---|---|---|---|---|---|---|---|
| 14 | **Bhaji**, cabbage and potato, *with butter* | 78.6 | 0.37 | 2.3 | 5.8 | 9.1 | 95 | 398 |
| 15 | *with vegetable oil* | 77.5 | 0.36 | 2.2 | 7.0 | 9.1 | 106 | 442 |
| 16 | cabbage and spinach | 64.8 | 0.59 | 3.6 | 17.1 | 7.8 | 198 | 817 |
| 17 | carrot, potato and pea, *with butter* | 75.6 | 0.39 | 2.4 | 6.6 | 12.2 | 115 | 480 |
| 18 | *with vegetable oil* | 74.5 | 0.38 | 2.4 | 8.0 | 12.2 | 127 | 530 |
| 19 | cauliflower | 65.5 | 0.65 | 4.0 | 20.5 | 4.0 | 214 | 883 |
| 20 | cauliflower and potato | 73.2 | 0.69 | 4.3 | 7.2 | 8.6 | 108 | 450 |
| 21 | cauliflower, potato and pea, *with butter* | 71.9 | 0.56 | 3.5 | 11.3 | 8.2 | 145 | 603 |
| 22 | *with vegetable oil* | 69.9 | 0.55 | 3.4 | 13.7 | 8.2 | 166 | 688 |
| 23 | cauliflower and vegetable | 73.7 | 0.64 | 3.9 | 9.4 | 5.3 | 117 | 486 |
| 24 | green bean | 79.0 | 0.35 | 2.2 | 7.6 | 5.0 | 91 | 379 |
| 25 | karela, *with butter ghee* | 81.4 | 0.29 | 1.8 | 10.6 | 1.2 | 103 | 426 |
| 26 | *with vegetable oil* | 81.4 | 0.29 | 1.8 | 10.6 | 1.2 | 103 | 426 |
| 27 | mushroom | 73.6 | 0.47 | 1.7 | 16.1 | 4.4 | 166 | 688 |
| 28 | mustard leaves | 81.0 | 0.49 | 3.0 | 6.2 | 4.7 | 85 | 356 |
| 29 | mustard leaves and spinach | 80.4 | 0.50 | 3.1 | 6.4 | 3.9 | 84 | 351 |

# Vegetable Dishes

Carbohydrate fractions, g per 100g food

| No. 15- | Food | Starch | Total sugars | Individual sugars | | | | | | Oligo-saccharides |
|---|---|---|---|---|---|---|---|---|---|---|
| | | | | Glucose | Fructose | Galactose | Sucrose | Maltose | Lactose | |
| 14 | **Bhaji**, cabbage and potato, *with butter* | 4.0 | 4.7 | 2.1 | 1.9 | 0 | 0.7 | 0 | Tr | 0.4 |
| 15 | *with vegetable oil* | 4.0 | 4.7 | 2.1 | 1.9 | 0 | 0.7 | 0 | 0 | 0.4 |
| 16 | cabbage and spinach | 3.3 | 3.8 | 1.6 | 1.4 | 0 | 0.9 | 0 | 0 | 0.6 |
| 17 | carrot, potato and pea, *with butter* | 5.2 | 6.1 | 1.9 | 1.6 | 0 | 2.6 | 0 | Tr | 1.0 |
| 18 | *with vegetable oil* | 5.2 | 6.1 | 1.9 | 1.6 | 0 | 2.6 | 0 | 0 | 1.0 |
| 19 | cauliflower | 0.4 | 3.2 | 1.6 | 1.3 | 0 | 0.3 | 0 | 0 | 0.4 |
| 20 | cauliflower and potato | 6.1 | 2.5 | 1.2 | 1.0 | 0 | 0.2 | 0 | 0 | 0.1 |
| 21 | cauliflower, potato and pea, *with butter* | 5.5 | 2.5 | 1.1 | 1.0 | 0 | 0.3 | 0 | Tr | 0.2 |
| 22 | *with vegetable oil* | 5.5 | 2.5 | 1.1 | 1.0 | 0 | 0.3 | 0 | 0 | 0.2 |
| 23 | cauliflower and vegetable | N | N | N | N | 0 | N | 0 | Tr | 0.5 |
| 24 | green bean | 0.9 | 4.1 | 1.6 | 2.1 | 0 | 0.4 | 0 | 0 | Tr |
| 25 | karela, *with butter ghee* | 0.9 | Tr | Tr | Tr | 0 | Tr | 0 | Tr | Tr |
| 26 | *with vegetable oil* | 0.9 | Tr | Tr | Tr | 0 | Tr | 0 | 0 | Tr |
| 27 | mushroom | 0.2 | 3.0 | 1.1 | 0.9 | 0 | 1.0 | 0 | 0 | 1.2 |
| 28 | mustard leaves | 3.2 | 1.3 | 0.6 | 0.5 | 0 | 0.1 | 0 | Tr | 0.2 |
| 29 | mustard leaves and spinach | 2.1 | 1.6 | 0.6 | 0.5 | 0 | 0.3 | 0 | Tr | 0.2 |

**Fibre fractions, phytic acid and fatty acids, g per 100g food**
**Cholesterol, mg per 100g food**

| No. 15- | Food | Dietary fibre, g Southgate method | Englyst method | Fibre fractions, g Cellulose | Non-cellulosic polysaccharide Soluble | Insoluble | Lignin | Phytic acid g | Fatty acids, g Satd | Mono- unsatd | Poly- unsatd | Cholesterol mg |
|---|---|---|---|---|---|---|---|---|---|---|---|---|
| 14 | **Bhaji,** cabbage and potato, *with* butter | (3.1) | 2.5 | 0.8 | 1.3 | 0.4 | 0.3 | 0.01 | 3.7 | 1.4 | 0.3 | 15 |
| 15 | *with vegetable oil* | (3.1) | 2.5 | 0.8 | 1.3 | 0.4 | 0.3 | 0.01 | 0.8 | 2.4 | 3.3 | 0 |
| 16 | cabbage and spinach | (4.2) | 3.0 | 1.0 | 1.3 | 0.7 | 0.4 | 0.06 | 1.8 | 5.9 | 8.2 | 0 |
| 17 | carrot, potato and pea, *with* butter | (3.7) | 3.0 | 1.3 | 1.4 | 0.3 | 0.1 | 0.02 | 4.2 | 1.5 | 0.5 | 17 |
| 18 | *with vegetable oil* | (3.7) | 3.0 | 1.3 | 1.4 | 0.3 | 0.1 | 0.02 | 0.9 | 2.7 | 3.9 | 0 |
| 19 | cauliflower | 2.2 | 2.0 | 0.5 | 1.0 | 0.5 | Tr | 0.07 | 2.2 | 7.0 | 9.9 | 0 |
| 20 | cauliflower and potato | 2.2 | 2.0 | 0.5 | 1.0 | 0.5 | Tr | 0.07 | 0.8 | 2.4 | 3.4 | 0 |
| 21 | cauliflower, potato and pea, *with butter* | 2.2 | 1.9 | 0.6 | 0.9 | 0.4 | Tr | 0.04 | 7.1 | 2.7 | 0.7 | 29 |
| 22 | *with vegetable oil* | 2.2 | 1.9 | 0.6 | 0.9 | 0.4 | Tr | 0.04 | 1.5 | 4.7 | 6.6 | 0 |
| 23 | cauliflower and vegetable | (2.2) | (2.0) | (0.5) | (0.9) | (0.6) | N | 0.06 | 5.6 | 2.2 | 0.8 | 22 |
| 24 | green bean | 2.9 | 2.1 | 0.8 | 0.9 | 0.5 | N | 0.02 | 0.7 | 2.6 | 3.3 | 0 |
| 25 | karela, *with butter ghee* | 3.3 | 2.4 | 0.8 | 1.1 | 0.5 | 0.3 | Tr | 6.6 | 2.5 | 0.5 | 27 |
| 26 | *with vegetable oil* | 3.3 | 2.4 | 0.8 | 1.1 | 0.5 | 0.3 | Tr | 1.1 | 3.6 | 5.0 | 0 |
| 27 | mushroom | 2.0 | 1.3 | 0.4 | 0.5 | 0.5 | Tr | 0.06 | 1.7 | 5.6 | 7.8 | 0 |
| 28 | mustard leaves | N | N | N | N | N | N | Tr | 3.9 | 1.4 | 0.3 | 16 |
| 29 | mustard leaves and spinach | N | N | N | N | N | N | Tr | 3.9 | 1.5 | 0.5 | 16 |

# Vegetable Dishes

**Inorganic constituents per 100g food**

| No. Food 15- | Na | K | Ca | Mg | P | Fe | Cu | Zn | Cl | Mn | Se | I |
|---|---|---|---|---|---|---|---|---|---|---|---|---|
| | | | | | | mg | | | | | | μg |
| | | | | | | | | | | | Se | I |
| 14 **Bhaji**, cabbage and potato, *with butter* | N | 360 | 48 | 12 | 53 | (0.8) | 0.05 | 0.4 | N | 0.2 | (1) | 5 |
| 15 *with vegetable oil* | N | 360 | 47 | 12 | 51 | (0.8) | 0.04 | 0.4 | N | 0.2 | (1) | 4 |
| 16 cabbage and spinach | 480 | 440 | 110 | 36 | 70 | 1.8 | 0.08 | 0.7 | 720 | 0.5 | (1) | 5 |
| 17 carrot, potato and pea, *with butter* | N | 280 | 28 | 12 | 48 | (0.7) | (0.05) | 0.4 | N | 0.2 | (1) | (6) |
| 18 *with vegetable oil* | N | 270 | 27 | 12 | 46 | (0.7) | (0.05) | 0.4 | N | 0.2 | (1) | (4) |
| 19 cauliflower | 420 | 430 | 27 | 22 | 73 | 0.9 | 0.04 | 0.7 | 670 | 0.3 | Tr | 3 |
| 20 cauliflower and potato | 250 | 500 | 41 | 31 | 81 | 2.3 | 0.09 | 0.8 | 400 | 0.4 | Tr | 2 |
| 21 cauliflower, potato and pea, *with butter* | 400 | 390 | 23 | 22 | 65 | 1.0 | (0.06) | 0.6 | 640 | 0.3 | Tr | 6 |
| 22 *with vegetable oil* | 300 | 390 | 21 | 22 | 62 | 1.0 | (0.05) | 0.6 | 500 | 0.3 | Tr | 3 |
| 23 cauliflower and vegetable | 520 | 460 | 50 | 30 | 79 | 2.4 | 0.08 | 0.7 | (810) | (0.4) | N | N |
| 24 green bean | 460 | 280 | 46 | 26 | 46 | 1.9 | 0.04 | 0.3 | 720 | N | N | N |
| 25 karela, *with butter ghee* | N | 340 | 29 | 33 | 51 | (2.2) | 0.26 | 0.5 | N | 0.4 | N | N |
| 26 *with vegetable oil* | N | 340 | 29 | 33 | 51 | (2.2) | 0.26 | 0.5 | N | 0.4 | N | N |
| 27 mushroom | 9 | 280 | 18 | 8 | 62 | 0.7 | 0.43 | 0.4 | 58 | 0.1 | 5 | 5 |
| 28 mustard leaves | 720 | 390 | 150 | 42 | 56 | 2.6 | N | N | N | N | N | N |
| 29 mustard leaves and spinach | 770 | 450 | 160 | 50 | 56 | 2.6 | N | N | N | N | N | N |

# Vegetable Dishes

| No. 15- | Food | Retinol µg | Carotene µg | Vitamin D µg | Vitamin E mg | Thiamin mg | Ribo-flavin mg | Niacin mg | Trypt 60 mg | Vitamin B6 mg | Vitamin B12 µg | Folate µg | Panto-thenate mg | Biotin µg | Vitamin C mg |
|---|---|---|---|---|---|---|---|---|---|---|---|---|---|---|---|
| 14 | **Bhaji**, cabbage and potato, *with butter* | 53 | 400 | Tr | 0.50 | 0.16 | 0.02 | 0.6 | 0.4 | 0.23 | Tr | 35 | 0.24 | 0.4 | 22 |
| 15 | *with vegetable oil* | 0 | 370 | 0 | 1.95 | 0.16 | 0.02 | 0.6 | 0.4 | 0.23 | 0 | 35 | 0.23 | 0.4 | 22 |
| 16 | cabbage and spinach | 0 | 1560 | 0 | 4.88 | 0.12 | 0.05 | 0.8 | 0.7 | 0.18 | 0 | 54 | (0.27) | (0.2) | 18 |
| 17 | carrot, potato and pea, *with butter* | 61 | 4530 | 0.1 | 0.71 | 0.15 | 0.03 | 0.7 | 0.4 | 0.20 | Tr | 14 | 0.24 | 0.7 | 6 |
| 18 | *with vegetable oil* | 0 | 4495 | 0 | 2.37 | 0.15 | 0.02 | 0.7 | 0.4 | 0.20 | 0 | 14 | 0.24 | 0.7 | 6 |
| 19 | cauliflower | 0 | 185 | 0 | 4.94 | 0.16 | 0.05 | 0.6 | 1.0 | 0.25 | 0 | 35 | 0.51 | 1.3 | 24 |
| 20 | cauliflower and potato | 0 | 220 | 0 | 1.63 | 0.18 | 0.05 | 0.7 | 1.0 | 0.29 | 0 | 33 | 0.53 | 1.1 | 20 |
| 21 | cauliflower, potato and pea, *with butter* | 105 | 195 | 0.1 | 0.50 | 0.16 | 0.04 | 0.6 | 0.8 | 0.26 | Tr | 29 | 0.42 | 0.9 | 17 |
| 22 | *with vegetable oil* | 0 | 140 | 0 | 3.34 | 0.16 | 0.04 | 0.6 | 0.8 | 0.26 | 0 | 29 | 0.41 | 0.9 | 17 |
| 23 | cauliflower and vegetable | 80 | 665 | 0.1 | (0.51) | 0.15 | 0.06 | 0.8 | 0.9 | (0.22) | Tr | 28 | (0.39) | (1.1) | 21 |
| 24 | green bean | 0 | 420 | 0 | 1.96 | 0.05 | 0.06 | 0.8 | 0.5 | 0.05 | 0 | (37) | 0.09 | 0.9 | 7 |
| 25 | karela, *with butter ghee* | 67 | 780 | 0.2 | N | 0.07 | 0.05 | 0.5 | 0.3 | N | Tr | 19 | N | N | 81 |
| 26 | *with vegetable oil* | 0 | 730 | 0 | N | 0.07 | 0.05 | 0.5 | 0.3 | N | 0 | 19 | N | N | 81 |
| 27 | mushroom | 0 | 105 | 0 | 4.00 | 0.10 | 0.14 | 1.7 | 0.3 | 0.16 | 0 | 16 | 0.92 | 5.6 | 2 |
| 28 | mustard leaves | 57 | 700 | Tr | N | 0.07 | 0.12 | 0.8 | N | N | Tr | N | 0.19 | N | 39 |
| 29 | mustard leaves and spinach | 57 | 1875 | Tr | N | 0.07 | 0.11 | 0.9 | N | N | Tr | N | (0.21) | N | 30 |

| No. | Food | Description and main data sources |
|-----|------|-----------------------------------|
| **15-** | | |
| 30 | **Bhaji,** okra, Bangladeshi, *with butter ghee* | Okra, onion and spices. Recipe adapted from Carlson and de Wet, 1991 |
| 31 | *with vegetable oil* | Okra, onion and spices. Recipe adapted from Carlson and de Wet 1991 |
| 32 | okra, Islami | Recipe from the Aga Khan Health Board for the United Kingdom. Very dry dish |
| 33 | pea | Pakistani dish. Recipe from Wharton et al, 1983 |
| 34 | potato, *with butter ghee* | Recipe from Wharton et al, 1983 |
| 35 | *with vegetable oil* | Recipe from Wharton et al, 1983 |
| 36 | potato and fenugreek leaves | Gujerati dish. Recipe from review of recipe collection |
| 37 | potato and green pepper | Punjabi dish. Potato, pepper, onion and tomato. Recipe from dietary survey records |
| 38 | potato and onion | Bangladeshi dish. Recipe adapted from Carlson and de Wet, 1991 |
| 39 | potato, onion and mushroom | Recipe from review of recipe collection |
| 40 | potato, spinach and cauliflower | |
| 41 | spinach | Recipe from Wharton et al, 1983 |
| 42 | spinach and potato | Recipe from review of recipe collection |
| 43 | turnip | Recipe from Wharton et al, 1983 |
| 44 | turnip and onion | Punjabi dish. Recipe from dietary survey records |
| 45 | vegetable, *with butter* | Recipe from Tan et al, 1985. Mixed vegetables |
| 46 | *with vegetable oil* | Recipe from Tan et al, 1985. Mixed vegetables |
| 47 | vegetable, Punjabi, *with butter* | Recipe from manufacturer. Mixed vegetables |
| 48 | *with vegetable oil* | Recipe from manufacturer. Mixed vegetables |

Composition of food per 100g

| No. 15- | Food | Water g | Total nitrogen g | Protein g | Fat g | Carbohydrate g | Energy value kcal | kJ |
|---|---|---|---|---|---|---|---|---|
| 30 | **Bhaji**, okra, Bangladeshi, *with butter ghee* | 77.7 | 0.39 | 2.5 | 6.4 | 7.6 | 95 | 395 |
| 31 | *with vegetable oil* | 77.7 | 0.39 | 2.5 | 6.4 | 7.6 | 95 | 395 |
| 32 | okra, Islami | 57.8 | 0.71 | 4.8 | 16.4 | 8.7 | 192 | 797 |
| 33 | pea | 44.3 | 0.79 | 4.9 | 34.0 | 9.0 | 356 | 1466 |
| 34 | potato, *with butter ghee* | 67.2 | 0.37 | 2.3 | 10.0 | 16.3 | 158 | 659 |
| 35 | *with vegetable oil* | 67.2 | 0.37 | 2.3 | 10.0 | 16.3 | 158 | 659 |
| 36 | potato and fenugreek leaves | 55.9 | 0.61 | 3.8 | 15.9 | 18.6 | 214 | 891 |
| 37 | potato and green pepper | 75.1 | 0.27 | 1.7 | 9.1 | 11.4 | 131 | 546 |
| 38 | potato and onion | 68.5 | 0.35 | 2.1 | 10.1 | 16.6 | 160 | 668 |
| 39 | potato, onion and mushroom | 65.2 | 0.39 | 2.0 | 17.5 | 12.0 | 208 | 864 |
| 40 | potato, spinach and cauliflower | 72.0 | 0.35 | 2.2 | 15.1 | 7.1 | 169 | 699 |
| 41 | spinach | 79.4 | 0.53 | 3.3 | 6.8 | 2.6 | 83 | 343 |
| 42 | spinach and potato | 62.6 | 0.59 | 3.7 | 14.1 | 13.4 | 191 | 793 |
| 43 | turnip | 70.3 | 0.20 | 1.3 | 16.6 | 6.3 | 175 | 726 |
| 44 | turnip and onion | 77.9 | 0.21 | 1.3 | 10.9 | 7.1 | 128 | 533 |
| 45 | vegetable, *with butter* | 68.0 | 0.35 | 2.2 | 15.2 | 10.1 | 183 | 758 |
| 46 | *with vegetable oil* | 65.2 | 0.34 | 2.1 | 18.5 | 10.1 | 212 | 878 |
| 47 | vegetable, Punjabi, *with butter* | 75.0 | 0.57 | 3.5 | 7.2 | 8.2 | 108 | 449 |
| 48 | *with vegetable oil* | 73.8 | 0.56 | 3.5 | 8.6 | 8.2 | 120 | 501 |

# Vegetable Dishes

**Carbohydrate fractions, g per 100g food**

| No. 15- | Food | Starch | Total sugars | Individual sugars | | | | | | Oligo-saccharides |
|---|---|---|---|---|---|---|---|---|---|---|
| | | | | Glucose | Fructose | Galactose | Sucrose | Maltose | Lactose | |
| 30 | **Bhaji**, okra, Bangladeshi, *with butter ghee* | 0.4 | 5.5 | 1.9 | 1.7 | 0 | 1.9 | 0 | Tr | 1.7 |
| 31 | *with vegetable oil* | 0.4 | 5.5 | 1.9 | 1.7 | 0 | 1.9 | 0 | 0 | 1.7 |
| 32 | okra, Islami | 0.9 | 6.6 | 2.1 | 2.2 | 0 | 2.1 | 0 | 0 | 1.2 |
| 33 | pea | 4.2 | 2.9 | 0.7 | 0.5 | 0 | 1.8 | 0 | Tr | 1.8 |
| 34 | potato, *with butter ghee* | 14.5 | 1.5 | 0.5 | 0.4 | 0 | 0.6 | 0 | Tr | 0.4 |
| 35 | *with vegetable oil* | 14.5 | 1.5 | 0.5 | 0.4 | 0 | 0.6 | 0 | 0 | 0.4 |
| 36 | potato and fenugreek leaves | 14.5 | 4.1 | 0.2 | 0.1 | 0 | 3.7 | 0 | 0 | 0 |
| 37 | potato and green pepper | 8.7 | 2.4 | 1.0 | 1.0 | 0 | 0.5 | 0 | 0 | 0.4 |
| 38 | potato and onion | 12.1 | 3.3 | 1.2 | 0.9 | 0 | 1.2 | 0 | Tr | 1.1 |
| 39 | potato, onion and mushroom | 7.2 | 3.5 | 1.3 | 1.0 | 0 | 1.2 | 0 | 0 | 1.3 |
| 40 | potato, spinach and cauliflower | 5.2 | 1.6 | 0.6 | 0.5 | 0 | 0.5 | 0 | Tr | 0.3 |
| 41 | spinach | 0.3 | 2.0 | 0.7 | 0.7 | 0 | 0.7 | 0 | Tr | 0.2 |
| 42 | spinach and potato | 11.8 | 1.6 | 0.5 | 0.5 | 0 | 0.6 | 0 | 0 | Tr |
| 43 | turnip | 0.6 | 5.4 | 2.7 | 2.0 | 0 | 0.8 | 0 | Tr | 0.3 |
| 44 | turnip and onion | 0.3 | 5.6 | 2.5 | 1.9 | 0 | 1.3 | 0 | 0 | 1.2 |
| 45 | vegetable, *with butter* | 5.0 | 4.6 | 1.8 | 1.5 | 0 | 1.4 | 0 | Tr | 0.4 |
| 46 | *with vegetable oil* | 5.0 | 4.6 | 1.8 | 1.5 | 0 | 1.4 | 0 | 0 | 0.4 |
| 47 | vegetable, Punjabi, *with butter* | 1.5 | 5.5 | 1.9 | 1.7 | 0 | 1.9 | 0 | Tr | 1.2 |
| 48 | *with vegetable oil* | 1.5 | 5.5 | 1.9 | 1.7 | 0 | 1.9 | 0 | 0 | 1.2 |

# Vegetable Dishes

Fibre fractions, phytic acid and fatty acids, g per 100g food
Cholesterol, mg per 100g food

| No. 15- | Food | Dietary fibre, g Southgate method | Dietary fibre, g Englyst method | Cellulose | Non-cellulosic polysaccharide Soluble | Non-cellulosic polysaccharide Insoluble | Lignin | Phytic acid g | Fatty acids, g Satd | Fatty acids, g Mono-unsatd | Fatty acids, g Poly-unsatd | Cholesterol mg |
|---|---|---|---|---|---|---|---|---|---|---|---|---|
| 30 | **Bhaji**, okra, Bangladeshi, *with butter ghee* | 3.6 | 3.2 | 1.0 | 1.9 | 0.4 | 0.3 | 0.02 | 3.9 | 1.4 | 0.4 | 15 |
| 31 | *with vegetable oil* | 3.6 | 3.2 | 1.0 | 1.9 | 0.4 | 0.3 | 0.02 | 0.8 | 2.1 | 3.0 | 0 |
| 32 | okra, *Islami* | 6.5 | 5.8 | 1.8 | 3.4 | 0.6 | 0.6 | 0.03 | 1.9 | 5.5 | 7.6 | 0 |
| 33 | pea | 3.2 | 3.2 | 1.8 | 1.0 | 0.5 | 0.1 | 0.01 | 21.9 | 8.1 | 1.5 | 92 |
| 34 | potato, *with butter ghee* | 1.6 | 1.4 | 0.4 | 0.7 | 0.1 | Tr | 0.01 | 6.4 | 2.4 | 0.4 | 26 |
| 35 | *with vegetable oil* | 1.6 | 1.4 | 0.4 | 0.7 | 0.1 | Tr | 0.01 | 1.0 | 3.5 | 4.7 | 0 |
| 36 | potato and fenugreek leaves | N | N | 0.6 | N | N | 0.1 | 0.02 | 1.6 | 5.6 | 7.3 | 0 |
| 37 | potato and green pepper | (1.9) | 1.5 | 0.5 | 0.8 | 0.2 | 0.1 | 0.01 | 1.0 | 3.1 | 4.3 | 0 |
| 38 | potato and onion | 1.8 | 1.6 | 0.5 | 0.9 | 0.2 | Tr | 0.02 | 6.5 | 2.3 | 0.4 | 27 |
| 39 | potato, onion and mushroom | 2.0 | 1.5 | 0.5 | 0.8 | 0.3 | Tr | 0.04 | 1.8 | 6.1 | 8.4 | 0 |
| 40 | potato, spinach and cauliflower | 1.8 | 1.4 | 0.5 | 0.7 | 0.2 | Tr | 0.02 | 9.7 | 3.6 | 0.7 | 40 |
| 41 | spinach | (4.3) | 2.4 | 1.0 | 0.9 | 0.5 | 0.1 | Tr | 3.9 | 1.5 | 0.7 | 16 |
| 42 | spinach and potato | (3.8) | 2.3 | 0.9 | 1.0 | 0.3 | 0.1 | 0.01 | 1.5 | 4.8 | 6.7 | 0 |
| 43 | turnip | 2.7 | 2.6 | 1.2 | 1.0 | 0.3 | Tr | 0.01 | 10.8 | 4.1 | 0.7 | 45 |
| 44 | turnip and onion | 2.3 | 2.2 | 0.9 | 1.0 | 0.3 | Tr | 0.02 | 1.2 | 3.8 | 5.2 | 0 |
| 45 | vegetable, *with butter* | 2.7 | 2.4 | 0.8 | 1.2 | 0.3 | 0.1 | 0.02 | 9.8 | 3.6 | 0.7 | 41 |
| 46 | *with vegetable oil* | 2.7 | 2.4 | 0.8 | 1.2 | 0.3 | 0.1 | 0.02 | 2.0 | 6.5 | 8.9 | 0 |
| 47 | vegetable, Punjabi, *with butter* | 4.0 | (3.2) | (1.4) | (1.4) | (0.5) | (0.1) | 0.03 | 4.4 | 1.7 | 0.6 | 17 |
| 48 | *with vegetable oil* | 4.0 | (3.2) | (1.4) | (1.4) | (0.5) | (0.1) | 0.03 | 1.0 | 2.9 | 4.1 | 0 |

| No. 15- | Food | Na | K | Ca | Mg | P | Fe | Cu | Zn | Cl | Mn | Se µg | I µg |
|---|---|---|---|---|---|---|---|---|---|---|---|---|---|
|  |  | mg | | | | | | | | | | | |
| 30 | **Bhaji**, okra, Bangladeshi, *with butter ghee* | 160 | 310 | 110 | 44 | 57 | 1.0 | 0.11 | 0.5 | 290 | N | (1) | N |
| 31 | *with vegetable oil* | 160 | 310 | 110 | 44 | 57 | 1.0 | 0.11 | 0.5 | 280 | N | (1) | N |
| 32 | okra, Islami | 480 | 590 | 230 | 100 | 110 | 2.5 | 0.24 | 1.0 | 790 | N | (2) | N |
| 33 | pea | 1040 | 290 | 34 | 31 | 100 | 2.3 | 0.07 | 0.8 | 1620 | 0.3 | (1) | 18 |
| 34 | potato, *with butter ghee* | 380 | 390 | 19 | 24 | 45 | 1.1 | 0.10 | 0.4 | 640 | 0.2 | 1 | 4 |
| 35 | *with vegetable oil* | 380 | 390 | 19 | 24 | 45 | 1.1 | 0.10 | 0.4 | 640 | 0.2 | 1 | 5 |
| 36 | potato and fenugreek leaves | 50 | 400 | 82 | 47 | 67 | 5.1 | 0.19 | 0.5 | 130 | N | N | N |
| 37 | potato and green pepper | N | 290 | 10 | 14 | 35 | (0.5) | 0.06 | 0.2 | N | 0.1 | 1 | 4 |
| 38 | potato and onion | 310 | 300 | 22 | 16 | 44 | 0.8 | 0.09 | 0.4 | 510 | 0.1 | 1 | 9 |
| 39 | potato, onion and mushroom | 10 | 330 | 19 | 13 | 50 | 0.7 | 0.19 | 0.4 | 62 | 0.2 | 3 | 5 |
| 40 | potato, spinach and cauliflower | 37 | 220 | 52 | 18 | 37 | 1.0 | 0.04 | 0.4 | 47 | 0.3 | (1) | 2 |
| 41 | spinach | 850 | 570 | 190 | 65 | 56 | 2.6 | 0.06 | 0.8 | 1180 | 0.7 | (1) | 6 |
| 42 | spinach and potato | 400 | 620 | 120 | 53 | 64 | 1.9 | 0.09 | 0.7 | 570 | 0.5 | (1) | 5 |
| 43 | turnip | 540 | 340 | 57 | 16 | 49 | 1.2 | 0.03 | 0.2 | 860 | 0.5 | (1) | N |
| 44 | turnip and onion | N | 260 | 42 | 7 | 42 | (0.3) | 0.03 | 0.2 | N | 0.1 | (1) | N |
| 45 | vegetable, *with butter* | 390 | 340 | 28 | 16 | 47 | 0.9 | 0.05 | 0.4 | 630 | 0.3 | (1) | (9) |
| 46 | *with vegetable oil* | 260 | 340 | 26 | 16 | 43 | 0.8 | 0.05 | 0.4 | 430 | 0.3 | (1) | (4) |
| 47 | vegetable, Punjabi, *with butter* | 450 | 340 | 41 | 23 | 68 | 1.5 | (0.06) | 0.6 | 710 | (0.3) | (1) | (5) |
| 48 | *with vegetable oil* | 390 | 340 | 40 | 23 | 66 | 1.5 | (0.06) | 0.6 | 620 | (0.3) | (1) | (3) |

## Vegetable Dishes

Vitamins per 100g food

| No. 15- | Food | Retinol μg | Carotene μg | Vitamin D μg | Vitamin E mg | Thiamin mg | Ribo-flavin mg | Niacin mg | Trypt 60 mg | Vitamin B6 mg | Vitamin B12 μg | Folate μg | Panto-thenate mg | Biotin μg | Vitamin C mg |
|---|---|---|---|---|---|---|---|---|---|---|---|---|---|---|---|
| 30 | **Bhaji,** okra, Bangladeshi, *with butter ghee* | 38 | 320 | 0.1 | N | 0.17 | 0.03 | 0.9 | 0.5 | 0.21 | Tr | 30 | 0.17 | N | 9 |
| 31 | *with vegetable oil* | 0 | 295 | 0 | N | 0.17 | 0.03 | 0.9 | 0.5 | 0.21 | 0 | 30 | 0.17 | N | 9 |
| 32 | okra, Islami | 0 | 800 | 0 | N | 0.27 | 0.07 | 1.5 | 0.7 | 0.31 | 0 | 59 | 0.32 | N | 15 |
| 33 | pea | 325 | 485 | 0.3 | 1.01 | 0.39 | (0.02) | (1.4) | 0.8 | 0.10 | Tr | 21 | (0.11) | 0.4 | 9 |
| 34 | potato, *with butter ghee* | 64 | 68 | 0.2 | 0.42 | 0.17 | 0.02 | 0.5 | 0.5 | 0.34 | Tr | 17 | (0.27) | (0.3) | 7 |
| 35 | *with vegetable oil* | 0 | 20 | 0 | 2.42 | 0.17 | 0.02 | 0.5 | 0.5 | 0.34 | 0 | 17 | (0.27) | (0.3) | 7 |
| 36 | potato and fenugreek leaves | 0 | 1870 | 0 | 3.57 | 0.18 | 0.10 | 0.9 | N | N | 0 | N | N | N | 18 |
| 37 | potato and green pepper | 0 | 180 | 0 | 2.62 | 0.12 | 0.01 | 0.5 | 0.4 | 0.30 | 0 | 17 | 0.22 | N | 26 |
| 38 | potato and onion | 98 | 63 | 0.1 | 0.44 | 0.16 | 0.01 | 0.8 | 0.5 | 0.26 | Tr | 12 | 0.34 | 0.7 | 7 |
| 39 | potato, onion and mushroom | 0 | 115 | 0 | 4.34 | 0.15 | 0.07 | 1.2 | 0.5 | 0.25 | 0 | 15 | 0.56 | 2.7 | 6 |
| 40 | potato, spinach and cauliflower | 97 | 1090 | 0.3 | 0.91 | 0.10 | 0.03 | 0.5 | 0.5 | 0.17 | Tr | (42) | 0.27 | 0.4 | 10 |
| 41 | spinach | 57 | 3915 | Tr | 1.97 | 0.07 | 0.08 | 1.1 | 0.9 | 0.16 | Tr | 80 | (0.24) | (0.1) | 15 |
| 42 | spinach and potato | 0 | 2500 | 0 | 4.40 | 0.16 | 0.07 | 1.0 | 0.9 | 0.34 | 0 | 63 | (0.35) | (0.2) | 14 |
| 43 | turnip | 110 | 115 | 0.3 | 0.57 | 0.05 | 0.01 | 0.5 | 0.3 | 0.09 | Tr | 8 | 0.18 | 0.2 | 10 |
| 44 | turnip and onion | 0 | N | 0 | 2.71 | 0.08 | 0.01 | 0.5 | 0.3 | 0.12 | 0 | 8 | 0.14 | 0.4 | 7 |
| 45 | vegetable, *with butter* | 145 | 2405 | 0.1 | 0.72 | 0.13 | 0.02 | 0.5 | 0.5 | 0.22 | Tr | 17 | 0.29 | N | 9 |
| 46 | *with vegetable oil* | 0 | 2330 | 0 | 4.69 | 0.13 | 0.02 | 0.5 | 0.5 | 0.22 | 0 | 17 | 0.28 | N | 9 |
| 47 | vegetable, Punjabi, *with butter* | 63 | 2155 | 0.1 | 0.72 | 0.16 | 0.05 | 1.0 | 0.7 | 0.18 | Tr | (31) | 0.28 | 1.1 | 12 |
| 48 | *with vegetable oil* | 0 | 2125 | 0 | 2.44 | 0.16 | 0.05 | 1.0 | 0.7 | 0.18 | 0 | (31) | 0.27 | 1.1 | 12 |

| No. 15- | Food | Description and main data sources |
|---------|------|-----------------------------------|
| 49 | **Broccoli in cheese sauce,** *made with whole milk* | Recipe from review of recipe collection |
| 50 | *made with semi-skimmed milk* | Recipe from review of recipe collection |
| 51 | *made with skimmed milk* | Recipe from review of recipe collection |
| 52 | **Bubble and squeak,** *fried in lard* | Fried cabbage and potato. Recipe from Wiles et al, 1980 |
| 53 | *fried in sunflower oil* | Fried cabbage and potato. Recipe from Wiles et al, 1980 |
| 54 | *fried in vegetable oil* | Fried cabbage and potato. Recipe from Wiles et al, 1980 |
| 55 | **Cabbage,** red, *cooked with apple* | Recipe from Wiles et al, 1980 |
| 56 | **Callaloo and cho cho** | West Indian dish. Spinach and cho cho. Recipe from a personal collection |
| 57 | **Callaloo and okra** | West Indian dish. Spinach and okra. Recipe from a personal collection |
| 58 | **Cannelloni,** spinach | Pasta tubes with spinach and ricotta filling. Recipe from dietary survey records |
| 59 | vegetable | Pasta tubes with mixed vegetable filling. Recipe from dietary survey records |
| 60 | **Casserole,** bean and mixed vegetable | Recipe adapted from a recipe from Leeds Polytechnic. Mixed beans and vegetables |
| 61 | bean and root vegetable | Recipe from dietary survey records. Mixed beans and vegetables |
| 62 | sweet potato and green banana | West Indian dish. Recipe from a personal collection |
| 63 | vegetable | Recipe from dietary survey records |

**Vegetable Dishes**

| No. 15- | Food | Water g | Total nitrogen g | Protein g | Fat g | Carbohydrate g | Energy value kcal | Energy value kJ |
|---|---|---|---|---|---|---|---|---|
| 49 | **Broccoli in cheese sauce,** | | | | | | | |
| | *made with whole milk* | 77.3 | 1.03 | 6.5 | 8.4 | 4.6 | 118 | 494 |
| 50 | *made with semi-skimmed milk* | 78.1 | 1.04 | 6.5 | 7.5 | 4.7 | 111 | 463 |
| 51 | *made with skimmed milk* | 78.6 | 1.04 | 6.5 | 6.9 | 4.7 | 106 | 442 |
| 52 | **Bubble and squeak,** *fried in lard* | 77.6 | 0.23 | 1.4 | 9.1 | 9.8 | 124 | 516 |
| 53 | *fried in sunflower oil* | 77.5 | 0.23 | 1.4 | 9.1 | 9.8 | 124 | 519 |
| 54 | *fried in vegetable oil* | 77.5 | 0.23 | 1.4 | 9.1 | 9.8 | 124 | 519 |
| 55 | **Cabbage,** red, *cooked with apple* | 84.5 | 0.15 | 0.9 | 2.8 | 8.0 | 59 | 247 |
| 56 | **Callaloo and cho cho** | 83.7 | 0.33 | 2.0 | 6.6 | 2.6 | 76 | 315 |
| 57 | **Callaloo and okra** | 85.1 | 0.28 | 1.8 | 6.4 | 2.2 | 72 | 300 |
| 58 | **Cannelloni,** spinach | 73.4 | 0.74 | 4.5 | 7.5 | 12.6 | 132 | 553 |
| 59 | vegetable | 72.1 | 0.70 | 4.3 | 9.0 | 12.7 | 145 | 608 |
| 60 | **Casserole,** bean and mixed vegetable | 83.4 | 0.45 | 2.8 | 0.7 | 7.7 | 46 | 195 |
| 61 | bean and root vegetable | 82.2 | 0.48 | 3.0 | 0.6 | 7.8 | 46 | 197 |
| 62 | sweet potato and green banana | 58.9 | 0.17 | 1.0 | 6.5 | 30.6 | 177 | 748 |
| 63 | vegetable | 84.2 | 0.35 | 2.2 | 0.4 | 10.6 | 52 | 221 |

27

## Carbohydrate fractions, g per 100g food

| No. 15- | Food | Starch | Total sugars | Individual sugars | | | | | | Oligo-saccharides |
|---|---|---|---|---|---|---|---|---|---|---|
| | | | | Glucose | Fructose | Galactose | Sucrose | Maltose | Lactose | |
| 49 | **Broccoli in cheese sauce**, made with whole milk | 2.0 | 2.5 | 0.2 | 0.3 | 0 | 0.1 | Tr | 1.8 | 0.1 |
| 50 | made with semi-skimmed milk | 2.0 | 2.5 | 0.2 | 0.3 | 0 | 0.1 | Tr | 1.9 | 0.1 |
| 51 | made with skimmed milk | 2.0 | 2.5 | 0.2 | 0.3 | 0 | 0.1 | Tr | 1.9 | 0.1 |
| 52 | **Bubble and squeak**, fried in lard | 8.4 | 1.4 | 0.6 | 0.5 | 0 | 0.3 | 0 | 0 | Tr |
| 53 | fried in sunflower oil | 8.4 | 1.4 | 0.6 | 0.5 | 0 | 0.3 | 0 | 0 | Tr |
| 54 | fried in vegetable oil | 8.4 | 1.4 | 0.6 | 0.5 | 0 | 0.3 | 0 | 0 | Tr |
| 55 | **Cabbage**, red, cooked with apple | 0.1 | 7.5 | 1.5 | 2.0 | 0 | 3.9 | 0 | 0 | 0.3 |
| 56 | **Callaloo and cho cho** | 0.2 | 2.1 | 0.8 | 0.8 | 0 | 0.5 | 0 | 0 | 0.2 |
| 57 | **Callaloo and okra** | 0.2 | 1.8 | 0.6 | 0.6 | 0 | 0.5 | 0 | 0 | 0.2 |
| 58 | **Cannelloni**, spinach | 10.4 | 2.2 | 0.1 | 0.1 | 0 | 0.1 | 0.1 | 1.7 | Tr |
| 59 | vegetable | 10.3 | 2.4 | 0.2 | 0.2 | 0 | 0.1 | 0.1 | 1.7 | 0.1 |
| 60 | **Casserole**, bean and mixed vegetable | 3.5 | 3.8 | 1.1 | 1.1 | 0 | 1.7 | 0 | 0 | 0.4 |
| 61 | bean and root vegetable | 4.5 | 2.9 | 1.1 | 1.0 | 0 | 0.9 | 0 | 0 | 0.4 |
| 62 | sweet potato and green banana | 11.2 | 19.4 | N | N | 0 | N | N | Tr | 0 |
| 63 | vegetable | 5.5 | 4.6 | 1.4 | 1.3 | 0 | 1.8 | 0 | 0 | 0.5 |

Fibre fractions, phytic acid and fatty acids, g per 100g food
Cholesterol, mg per 100g food

| No. 15- | Food | Dietary fibre, g Southgate method | Englyst method | Cellulose | Non-cellulosic polysaccharide Soluble | Insoluble | Lignin | Phytic acid g | Fatty acids, g Satd | Mono- unsatd | Poly- unsatd | Cholesterol mg |
|---|---|---|---|---|---|---|---|---|---|---|---|---|
| 49 | **Broccoli in cheese sauce,** *made with whole milk* | (1.5) | 1.5 | 0.6 | 0.7 | 0.3 | Tr | Tr | 4.3 | 2.4 | 1.0 | 21 |
| 50 | *made with semi-skimmed milk* | (1.5) | 1.5 | 0.6 | 0.7 | 0.3 | Tr | Tr | 3.8 | 2.2 | 1.0 | 18 |
| 51 | *made with skimmed milk* | (1.5) | 1.5 | 0.6 | 0.7 | 0.3 | Tr | Tr | 3.4 | 2.0 | 1.0 | 16 |
| 52 | **Bubble and squeak,** *fried in lard* | (1.9) | 1.5 | 0.5 | 0.8 | 0.3 | 0.2 | Tr | 3.7 | 3.9 | 0.9 | 8 |
| 53 | *fried in sunflower oil* | (1.9) | 1.5 | 0.5 | 0.8 | 0.3 | 0.2 | Tr | 1.1 | 1.9 | 5.7 | 0 |
| 54 | *fried in vegetable oil* | (1.9) | 1.5 | 0.5 | 0.8 | 0.3 | 0.2 | Tr | 1.0 | 3.2 | 4.3 | 0 |
| 55 | **Cabbage,** red, cooked with *apple* | 2.7 | 2.1 | 0.8 | 1.0 | 0.4 | 0.2 | Tr | 0.7 | 0.8 | 1.2 | Tr |
| 56 | **Callaloo and cho cho** | (2.7) | 1.6 | 0.7 | 0.6 | 0.3 | 0.1 | 0.01 | 3.0 | 1.7 | 1.4 | 11 |
| 57 | **Callaloo and okra** | (2.5) | 1.6 | 0.6 | 0.7 | 0.3 | 0.1 | 0.01 | 2.9 | 1.7 | 1.3 | 11 |
| 58 | **Cannelloni,** spinach | 1.2 | 0.8 | 0.2 | 0.4 | 0.3 | 0.1 | N | 2.3 | 2.3 | 2.3 | 11 |
| 59 | *vegetable* | 1.0 | 0.7 | 0.1 | 0.3 | 0.2 | 0.1 | N | 3.4 | 2.7 | 2.3 | 15 |
| 60 | **Casserole,** bean and mixed *vegetable* | (2.2) | 1.9 | 0.6 | 0.8 | 0.5 | 0.1 | (0.07) | 0.1 | 0.1 | 0.3 | 0 |
| 61 | *bean and root vegetable* | (2.1) | 2.0 | 0.6 | 0.9 | 0.5 | N | 0.07 | 0.1 | 0.1 | 0.2 | 0 |
| 62 | *sweet potato and green banana* | 1.7 | 1.7 | 0.6 | 0.9 | 0.3 | N | 0.01 | 4.2 | 1.6 | 0.3 | 17 |
| 63 | *vegetable* | 2.5 | 2.1 | 0.7 | 1.0 | 0.4 | 0.1 | (0.01) | 0.1 | 0.1 | 0.1 | 0 |

| No. 15- | Food | mg | | | | | | | | | | µg | |
|---|---|---|---|---|---|---|---|---|---|---|---|---|---|
| | | Na | K | Ca | Mg | P | Fe | Cu | Zn | Cl | Mn | Se | I |
| 49 | **Broccoli in cheese sauce,** made with whole milk | 230 | 170 | 160 | 16 | 130 | 0.7 | 0.02 | 0.7 | (370) | 0.1 | 2 | 13 |
| 50 | made with semi-skimmed milk | 230 | 170 | 160 | 16 | 130 | 0.7 | 0.02 | 0.7 | (370) | 0.1 | 2 | 13 |
| 51 | made with skimmed milk | 230 | 170 | 160 | 16 | 130 | 0.7 | 0.03 | 0.7 | (370) | 0.1 | 2 | 13 |
| 52 | **Bubble and squeak,** fried in lard | 7 | 200 | 19 | 9 | 28 | 0.4 | 0.04 | 0.2 | 27 | 0.2 | 2 | 3 |
| 53 | fried in sunflower oil | 7 | 200 | 19 | 9 | 28 | 0.4 | 0.04 | 0.2 | 27 | 0.2 | 2 | 3 |
| 54 | fried in vegetable oil | 7 | 200 | 19 | 9 | 28 | 0.4 | 0.04 | 0.2 | 27 | 0.2 | 2 | 4 |
| 55 | **Cabbage,** red, cooked with apple | 32 | 210 | 46 | 8 | 31 | 0.4 | 0.02 | 0.1 | 74 | 0.2 | (2) | (2) |
| 56 | **Callaloo and cho cho** | 360 | 370 | 110 | 38 | 38 | 1.6 | (0.04) | 0.5 | 510 | (0.4) | (1) | (4) |
| 57 | **Callaloo and okra** | 470 | 300 | 96 | 35 | 33 | 1.1 | 0.04 | 0.4 | 690 | N | (1) | N |
| 58 | **Cannelloni,** spinach | 280 | 110 | N | 17 | 84 | 0.6 | 0.05 | 0.6 | N | 0.2 | N | N |
| 59 | vegetable | 280 | 100 | N | 13 | 83 | 0.4 | 0.05 | 0.6 | N | 0.1 | N | N |
| 60 | **Casserole,** bean and mixed vegetable | 130 | 220 | 19 | 15 | 66 | 0.7 | 0.06 | 0.3 | 220 | 0.2 | (1) | (1) |
| 61 | bean and root vegetable | 93 | 270 | 23 | 17 | 62 | 0.7 | 0.08 | 0.3 | 190 | 0.2 | 1 | (2) |
| 62 | sweet potato and green banana | 260 | 300 | 16 | 36 | 41 | 0.5 | 0.12 | 0.3 | 420 | 0.4 | (1) | N |
| 63 | vegetable | 74 | 320 | 23 | 16 | 56 | 0.6 | 0.06 | 0.3 | 140 | 0.2 | 1 | (3) |

| No. 15- | Food | Retinol µg | Carotene µg | Vitamin D µg | Vitamin E mg | Thiamin mg | Ribo-flavin mg | Niacin mg | Trypt/60 mg | Vitamin B6 mg | Vitamin B12 µg | Folate µg | Panto-thenate mg | Biotin µg | Vitamin C mg |
|---|---|---|---|---|---|---|---|---|---|---|---|---|---|---|---|
| 49 | **Broccoli in cheese sauce,** | | | | | | | | | | | | | | |
| | *made with whole milk* | 80 | 345 | 0.3 | (0.99) | 0.04 | 0.15 | 0.5 | 1.4 | 0.10 | 0.3 | 43 | N | N | 27 |
| 50 | *made with semi-skimmed milk* | 68 | 340 | 0.3 | (0.97) | 0.04 | 0.15 | 0.5 | 1.5 | 0.10 | 0.3 | 43 | N | N | 27 |
| 51 | *made with skimmed milk* | 60 | 340 | 0.3 | (0.96) | 0.04 | 0.15 | 0.5 | 1.5 | 0.10 | 0.3 | 43 | N | N | 27 |
| 52 | **Bubble and squeak,** *fried in lard* | Tr | 105 | N | 0.13 | 0.11 | 0.01 | 0.4 | 0.3 | 0.16 | Tr | 12 | 0.27 | 0.2 | 9 |
| 53 | *fried in sunflower oil* | 0 | 105 | 0 | 4.51 | 0.11 | 0.01 | 0.4 | 0.3 | 0.16 | 0 | 12 | 0.27 | 0.2 | 9 |
| 54 | *fried in vegetable oil* | 0 | 105 | 0 | 2.28 | 0.11 | 0.01 | 0.4 | 0.3 | 0.16 | 0 | 12 | 0.27 | 0.2 | 9 |
| 55 | **Cabbage,** *red, cooked with apple* | 25 | 37 | 0.3 | 0.21 | 0.02 | 0.01 | 0.3 | 0.2 | 0.07 | 0 | 14 | 0.19 | 0.3 | 21 |
| 56 | **Callaloo and cho cho** | 40 | 2200 | Tr | 1.69 | 0.05 | 0.05 | 0.8 | 0.5 | N | 0 | N | (0.20) | N | 10 |
| 57 | **Callaloo and okra** | 39 | 1690 | Tr | (1.47) | 0.06 | 0.04 | 0.7 | 0.4 | 0.11 | 0 | 39 | (0.15) | N | 8 |
| 58 | **Cannelloni,** *spinach* | 41 | 645 | N | (1.32) | 0.03 | 0.07 | 0.4 | 1.0 | 0.04 | 0.2 | (9) | (0.14) | (0.6) | 1 |
| 59 | *vegetable* | 57 | 315 | N | 1.20 | 0.03 | 0.07 | 0.3 | 1.0 | 0.04 | 0.2 | (4) | (0.13) | (0.7) | 1 |
| 60 | **Casserole,** *bean and mixed vegetable* | 0 | 1700 | 0 | (0.47) | 0.08 | 0.13 | 1.0 | 0.5 | 0.12 | 0 | 18 | (0.13) | N | 13 |
| 61 | *bean and root vegetable* | 0 | 770 | 0 | 0.56 | 0.10 | 0.08 | 0.8 | 0.5 | 0.13 | 0 | 17 | 0.16 | 2.4 | 4 |
| 62 | *sweet potato and green banana* | 62 | 2180 | 0.1 | 2.53 | 0.05 | 0.03 | 0.3 | 0.2 | 0.11 | Tr | 14 | N | N | 9 |
| 63 | *vegetable* | 0 | 995 | 0 | 0.78 | 0.14 | 0.08 | 1.1 | 0.4 | 0.18 | 0 | 20 | 0.23 | 0.7 | 7 |

| No. Food 15- | Description and main data sources |
|---|---|
| 64 **Cauliflower cheese,** *made with whole milk* | Recipe from Holland et al, 1991b |
| 65 *made with semi-skimmed milk* | Recipe from Holland et al, 1991b |
| 66 *made with skimmed milk* | Recipe from Holland et al, 1991b |
| 67 **Cauliflower in white sauce,** *made with whole milk* | Recipe from Wiles et al, 1980 |
| 68 *made with semi-skimmed milk* | Recipe from Wiles et al, 1980 |
| 69 *made with skimmed milk* | Recipe from Wiles et al, 1980 |
| 70 **Cauliflower with onions and chilli pepper** | Bangladeshi dish. Recipe from a personal collection |
| 71 **Chilli,** bean and lentil | Recipe from Waltham Forest College |
| 72 Quorn | Recipe from Leeds Polytechnic |
| 73 vegetable | Recipe from dietary survey records |
| 74 vegetable, retail | 6 samples, 5 brands; cooked in conventional and microwave ovens according to packet directions |
| 75 **Cho cho fritters,** *fried in vegetable oil* | West Indian dish. Recipe from review of recipe collection |
| 76 **Coco fritters,** *fried in vegetable oil* | West Indian dish. Recipe from review of recipe collection |

# Vegetable Dishes

| No. Food 15- | Water g | Total nitrogen g | Protein g | Fat g | Carbohydrate g | Energy value kcal | kJ |
|---|---|---|---|---|---|---|---|
| 64 **Cauliflower cheese**, *made with whole milk* | 78.5 | 0.95 | 6.0 | 7.0 | 5.2 | 105 | 440 |
| 65 *made with semi-skimmed milk* | 79.0 | 0.95 | 6.0 | 6.4 | 5.2 | 100 | 420 |
| 66 *made with skimmed milk* | 79.3 | 0.95 | 6.0 | 6.0 | 5.2 | 97 | 406 |
| 67 **Cauliflower in white sauce**, *made with whole milk* | 85.0 | 0.53 | 3.3 | 4.0 | 5.0 | 68 | 284 |
| 68 *made with semi-skimmed milk* | 85.7 | 0.54 | 3.3 | 3.2 | 5.1 | 61 | 256 |
| 69 *made with skimmed milk* | 86.2 | 0.54 | 3.3 | 2.7 | 5.1 | 56 | 236 |
| 70 **Cauliflower with onions and chilli pepper** | 87.0 | 0.44 | 2.7 | 3.4 | 3.4 | 54 | 226 |
| 71 **Chilli**, bean and lentil | 72.6 | 0.84 | 5.2 | 2.6 | 13.1 | 91 | 383 |
| 72 **Quorn** | 79.7 | 0.76 | 4.7 | 4.2 | 6.9 | 81 | 339 |
| 73 **vegetable** | 82.3 | 0.49 | 3.0 | 0.6 | 10.8 | 57 | 242 |
| 74 **vegetable**, retail | 79.8 | 0.64 | 4.0 | 2.1 | 9.4 | 70 | 296 |
| 75 **Cho cho fritters**, *fried in vegetable oil* | 64.5 | 0.76 | 4.6 | 13.0 | 16.0 | 194 | 813 |
| 76 **Coco fritters**, *fried in vegetable oil* | 42.1 | 0.76 | 4.7 | 22.8 | 26.0 | 321 | 1338 |

Carbohydrate fractions, g per 100g food

| No. Food 15- | Starch | Total sugars | Individual sugars | | | | | | Oligo-saccharides |
|---|---|---|---|---|---|---|---|---|---|
| | | | Glucose | Fructose | Galactose | Sucrose | Maltose | Lactose | |
| 64 **Cauliflower cheese**, *made with whole milk* | 2.1 | 3.0 | 0.9 | 0.8 | 0 | 0.1 | Tr | 1.2 | 0.1 |
| 65 *made with semi-skimmed milk* | 2.1 | 3.0 | 0.9 | 0.8 | 0 | 0.1 | Tr | 1.3 | 0.1 |
| 66 *made with skimmed milk* | 2.1 | 3.0 | 0.9 | 0.8 | 0 | 0.1 | Tr | 1.3 | 0.1 |
| 67 **Cauliflower in white sauce,** *made with whole milk* | 2.0 | 3.0 | 0.6 | 0.6 | 0 | 0.1 | Tr | 1.7 | 0.1 |
| 68 *made with semi-skimmed milk* | 2.0 | 3.0 | 0.6 | 0.6 | 0 | 0.1 | Tr | 1.8 | 0.1 |
| 69 *made with skimmed milk* | 2.0 | 3.0 | 0.6 | 0.6 | 0 | 0.1 | Tr | 1.8 | 0.1 |
| 70 **Cauliflower with onions and chilli pepper** | 0.2 | 2.7 | 1.2 | 1.0 | 0 | 0.5 | 0 | 0 | 0.6 |
| 71 **Chilli,** bean and lentil | 7.9 | 4.3 | 1.3 | 1.4 | 0 | 1.6 | 0 | 0 | 0.8 |
| 72 Quorn | 2.3 | 3.9 | 1.1 | 1.2 | 0 | 1.3 | Tr | 0.2 | 0.7 |
| 73 vegetable | 6.1 | 4.1 | 0.9 | 0.9 | 0 | 2.3 | 0 | 0 | 0.6 |
| 74 vegetable, retail | 7.0 | 2.4 | 0.8 | 0.9 | 0 | 0.7 | 0 | 0 | Tr |
| 75 **Cho cho fritters,** *fried in vegetable oil* | 12.8 | 3.1 | 1.0 | 1.1 | 0 | Tr | Tr | 0.9 | 0 |
| 76 **Coco fritters,** *fried in vegetable oil* | 25.2 | 0.9 | 0.2 | 0.2 | 0 | 0.5 | Tr | 0 | Tr |

Fibre fractions, phytic acid and fatty acids, g per 100g food
Cholesterol, mg per 100g food

| No. 15- | Food | Dietary fibre, g | | Fibre fractions, g | | | | Phytic acid g | Fatty acids, g | | | Cholesterol mg |
|---|---|---|---|---|---|---|---|---|---|---|---|---|
| | | Southgate method | Englyst method | Cellulose | Non-cellulosic polysaccharide | | Lignin | | Satd | Mono-unsatd | Poly-unsatd | |
| | | | | | Soluble | Insoluble | | | | | | | |
| 64 | **Cauliflower cheese**, *made with whole milk* | 1.4 | 1.3 | 0.3 | 0.6 | 0.4 | Tr | 0.04 | 3.4 | 1.9 | 1.3 | 13 |
| 65 | *made with semi-skimmed milk* | 1.4 | 1.3 | 0.3 | 0.6 | 0.4 | Tr | 0.04 | 3.0 | 1.7 | 1.3 | 11 |
| 66 | *made with skimmed milk* | 1.4 | 1.3 | 0.3 | 0.6 | 0.4 | Tr | 0.04 | 2.8 | 1.6 | 1.3 | 10 |
| 67 | **Cauliflower in white sauce,** *made with whole milk* | 1.2 | 1.1 | 0.4 | 0.5 | 0.2 | Tr | 0.04 | 1.6 | 1.2 | 0.9 | 8 |
| 68 | *made with semi-skimmed milk* | 1.2 | 1.1 | 0.4 | 0.5 | 0.2 | Tr | 0.04 | 1.1 | 1.0 | 0.9 | 6 |
| 69 | *made with skimmed milk* | 1.2 | 1.1 | 0.4 | 0.5 | 0.2 | Tr | 0.04 | 0.8 | 0.8 | 0.9 | 4 |
| 70 | **Cauliflower with onions and chilli pepper** | 1.6 | 1.6 | 0.6 | 0.8 | 0.3 | Tr | 0.05 | 0.4 | 1.0 | 1.7 | 0 |
| 71 | **Chilli,** bean and lentil | (4.6) | 3.6 | 1.1 | 1.6 | 1.0 | N | (0.14) | 0.3 | 0.7 | 1.1 | 0 |
| 72 | Quorn | (2.9) | 2.5 | N | N | N | Tr | 0.03 | 0.5 | 1.3 | 1.9 | 0 |
| 73 | vegetable | (3.4) | 2.6 | 0.7 | 1.2 | 0.7 | N | (0.07) | 0.1 | 0.1 | 0.2 | 0 |
| 74 | vegetable, retail | N | N | N | N | N | N | N | (0.2) | (0.7) | (1.0) | 0 |
| 75 | **Cho cho fritters,** *fried in vegetable oil* | (1.3) | 1.2 | 0.3 | 0.6 | 0.3 | 0.1 | 0.02 | 2.8 | 4.5 | 4.6 | 69 |
| 76 | **Coco fritters,** *fried in vegetable oil* | 2.7 | 1.9 | 0.4 | 1.1 | 0.4 | 0.1 | 0.01 | 2.8 | 8.2 | 10.1 | 81 |

| No. 15- | Food | mg | | | | | | | | | | µg | |
|---|---|---|---|---|---|---|---|---|---|---|---|---|---|
| | | Na | K | Ca | Mg | P | Fe | Cu | Zn | Cl | Mn | Se | I |
| 64 | **Cauliflower cheese**, *made with whole milk* | 200 | 310 | 120 | 17 | 120 | 0.6 | 0.03 | 0.7 | 320 | 0.2 | 2 | 9 |
| 65 | *made with semi-skimmed milk* | 200 | 310 | 120 | 17 | 120 | 0.6 | 0.03 | 0.8 | 320 | 0.2 | 2 | (9) |
| 66 | *made with skimmed milk* | 200 | 310 | 120 | 18 | 120 | 0.6 | 0.03 | 0.7 | 320 | 0.2 | 2 | (9) |
| 67 | **Cauliflower in white sauce**, *made with whole milk* | 140 | 130 | 55 | 13 | 70 | 0.3 | 0.02 | 0.4 | 220 | 0.2 | Tr | 6 |
| 68 | *made with semi-skimmed milk* | 140 | 140 | 57 | 13 | 71 | 0.3 | 0.02 | 0.4 | 220 | 0.2 | Tr | 6 |
| 69 | *made with skimmed milk* | 140 | 140 | 57 | 13 | 70 | 0.4 | 0.02 | 0.4 | 220 | 0.2 | Tr | 6 |
| 70 | **Cauliflower with onions and chilli pepper** | 4 | 140 | 19 | 11 | 50 | 0.4 | 0.03 | 0.4 | 17 | 0.2 | Tr | 1 |
| 71 | **Chilli**, bean and lentil | 320 | 370 | 44 | 29 | 94 | 2.0 | 0.11 | 0.7 | 530 | 0.3 | 10 | N |
| 72 | Quorn | 130 | (180) | (17) | (10) | (31) | (0.5) | (0.05) | (0.2) | (120) | (0.1) | (1) | N |
| 73 | vegetable | 200 | 280 | 23 | 21 | 59 | 1.0 | 0.08 | 0.4 | 330 | 0.2 | 2 | N |
| 74 | vegetable, retail | 420 | 260 | 39 | 23 | 71 | 1.3 | 0.11 | 0.4 | 660 | 0.3 | N | N |
| 75 | **Cho cho fritters**, *fried in vegetable oil* | 370 | 170 | 63 | 20 | 83 | 0.9 | N | 0.5 | 600 | N | N | N |
| 76 | **Coco fritters**, *fried in vegetable oil* | 600 | 280 | 58 | 29 | 200 | 1.2 | 0.19 | 1.2 | 710 | 0.4 | N | N |

| No. 15- | Food | Retinol µg | Carotene µg | Vitamin D µg | Vitamin E mg | Thiamin mg | Ribo-flavin mg | Niacin mg | Trypt 60 mg | Vitamin B6 mg | Vitamin B12 µg | Folate µg | Panto-thenate mg | Biotin µg | Vitamin C mg |
|---|---|---|---|---|---|---|---|---|---|---|---|---|---|---|---|
| 64 | **Cauliflower cheese,** *made with whole milk* | 64 | 79 | 0.2 | 0.23 | 0.09 | 0.11 | 0.4 | 1.4 | 0.14 | 0.2 | 29 | 0.43 | 1.6 | 16 |
| 65 | *made with semi-skimmed milk* | 56 | 76 | 0.2 | 0.22 | 0.09 | 0.11 | 0.4 | 1.4 | 0.14 | 0.2 | 29 | 0.42 | 1.6 | 16 |
| 66 | *made with skimmed milk* | 51 | 74 | 0.2 | 0.21 | 0.09 | 0.11 | 0.4 | 1.4 | 0.14 | 0.2 | 29 | 0.42 | 1.6 | 16 |
| 67 | **Cauliflower in white sauce,** *made with whole milk* | 37 | 65 | 0.2 | 0.31 | 0.06 | 0.09 | 0.3 | 0.8 | 0.12 | 0.1 | 35 | 0.39 | 1.4 | 18 |
| 68 | *made with semi-skimmed milk* | 26 | 61 | 0.2 | 0.29 | 0.06 | 0.09 | 0.3 | 0.8 | 0.12 | 0.1 | 35 | 0.38 | 1.4 | 18 |
| 69 | *made with skimmed milk* | 19 | 58 | 0.2 | 0.28 | 0.06 | 0.09 | 0.3 | 0.8 | 0.12 | 0.1 | 35 | 0.38 | 1.4 | 18 |
| 70 | **Cauliflower with onions and chilli pepper** | 0 | 55 | 0 | 0.79 | 0.08 | 0.03 | 0.5 | 0.7 | 0.16 | 0 | 44 | 0.37 | 1.0 | 24 |
| 71 | **Chilli,** bean and lentil | 0 | 1000 | 0 | 1.28 | 0.12 | 0.06 | 0.9 | 0.8 | 0.17 | 0 | 8 | 0.15 | N | 10 |
| 72 | Quorn | 0 | 595 | 0 | 1.33 | 7.93 | 0.05 | 0.6 | N | 0.10 | 0.1 | 5 | 0.13 | (2.5) | 10 |
| 73 | vegetable | 0 | 1170 | 0 | 0.54 | 0.08 | 0.02 | 0.6 | 0.5 | 0.10 | 0 | 14 | 0.16 | N | 7 |
| 74 | vegetable, retail | 0 | N | 0 | N | 0.08 | 0.05 | 0.6 | 0.6 | 0.16 | 0 | 35 | N | N | 10 |
| 75 | **Cho cho fritters,** *fried in vegetable oil* | 54 | 42 | 0.3 | 2.42 | 0.07 | 0.13 | 0.6 | 1.1 | N | 0.5 | N | 0.70 | N | 7 |
| 76 | **Coco fritters,** *fried in vegetable oil* | 40 | 24 | 0.4 | 5.14 | 0.08 | 0.12 | 0.7 | 1.3 | 0.07 | 0.5 | N | N | N | 6 |

| No. 15- | Food | Description and main data sources |
|---|---|---|
| 77 | **Coleslaw**, with mayonnaise, retail | Recipe from average of manufacturers proportions |
| 78 | with reduced calorie dressing, retail | Recipe from average of manufacturers proportions |
| 79 | with vinaigrette, retail | Recipe from manufacturer |
| 80 | **Coo-coo** | West Indian dish. Cornmeal with okra. Recipe from a personal collection |
| 81 | **Corn fritters**, *fried in vegetable oil* | West Indian dish. Sweetcorn-based. Recipe from review of recipe collection |
| 82 | **Corn pudding** | West Indian dish. Baked sweetcorn pudding. Recipe from a personal collection |
| 83 | **Courgettes with eggs** | Greek dish. Recipe from a personal collection |
| 84 | **Crumble**, vegetable in milk base | Cauliflower and sweetcorn crumble. Recipe from review of recipe collection |
| 85 | vegetable in milk base, wholemeal | Cauliflower and sweetcorn crumble. Recipe from review of recipe collection |
| 86 | vegetable in tomato base | Assorted vegetable crumble. Recipe from review of recipe collection |
| 87 | vegetable in tomato base, wholemeal | Assorted vegetable crumble. Recipe from review of recipe collection |
| 88 | **Curry**, almond | Recipe from Wharton et al, 1983 |
| 89 | aubergine | Aubergine and potato in masala sauce. Recipe from review of recipe collection |
| 90 | black gram dahl | Recipe from Tan et al, 1985. Split black gram |
| 91 | black gram, whole, Bengali | Recipe from Tan et al, 1985. Thin consistency |
| 92 | black gram, whole, Gujerati | Recipe from a personal collection. Thick consistency |

**Composition of food per 100g**

| No. Food 15- | Water g | Total nitrogen g | Protein g | Fat g | Carbohydrate g | Energy value kcal | Energy value kJ |
|---|---|---|---|---|---|---|---|
| 77 **Coleslaw**, with mayonnaise, retail | 65.6 | 0.20 | 1.2 | 26.4 | 4.2 | 258 | 939 |
| 78 with reduced calorie dressing, retail | 86.2 | 0.14 | 0.9 | 4.5 | 6.1 | 67 | 280 |
| 79 with vinaigrette, retail | 80.2 | 0.18 | 1.1 | 4.0 | 12.4 | 87 | 364 |
| 80 **Coo-coo** | 76.9 | 0.38 | 2.4 | 1.8 | 17.6 | 98 | 409 |
| 81 **Corn fritters**, *fried in vegetable oil* | 45.9 | 0.98 | 5.9 | 15.5 | 32.9 | 286 | 1199 |
| 82 **Corn pudding** | 77.3 | 0.75 | 4.7 | 3.9 | 14.7 | 109 | 460 |
| 83 **Courgettes with eggs** | 81.2 | 0.63 | 3.9 | 10.1 | 2.2 | 114 | 472 |
| 84 **Crumble**, vegetable in milk base | 63.5 | 0.87 | 5.3 | 8.9 | 21.8 | 183 | 768 |
| 85 vegetable in milk base, wholemeal | 63.5 | 0.96 | 5.9 | 9.1 | 19.5 | 178 | 745 |
| 86 vegetable in tomato base | 65.1 | 0.56 | 3.3 | 10.3 | 18.9 | 177 | 740 |
| 87 vegetable in tomato base, wholemeal | 65.1 | 0.66 | 3.9 | 10.5 | 16.4 | 171 | 716 |
| 88 **Curry**, almond | 49.7 | 1.02 | 5.6 | 33.7 | 3.3 | 327 | 1352 |
| 89 aubergine | 78.3 | 0.23 | 1.4 | 10.1 | 6.2 | 118 | 492 |
| 90 black gram dahl | 82.1 | 0.68 | 4.2 | 3.4 | 7.0 | 74 | 310 |
| 91 black gram, whole, Bengali | 79.5 | 0.70 | 4.4 | 2.8 | 8.6 | 76 | 320 |
| 92 black gram, whole, Gujerati | 60.3 | 1.34 | 8.4 | 6.6 | 15.0 | 151 | 634 |

| No. 15- | Food | Starch | Total sugars | Glucose | Fructose | Galactose | Sucrose | Maltose | Lactose | Oligo-saccharides |
|---|---|---|---|---|---|---|---|---|---|---|
| 77 | **Coleslaw**, with mayonnaise, retail | 0.2 | 3.9 | 1.5 | 1.4 | 0 | 1.0 | 0 | 0 | Tr |
| 78 | with reduced calorie dressing, retail | 0.1 | 6.0 | 1.5 | 1.4 | 0 | 3.1 | 0 | 0 | 0.1 |
| 79 | with vinaigrette, retail | 0.1 | 12.2 | 1.8 | 1.6 | 0 | 8.6 | 0 | 0 | 0.2 |
| 80 | **Coo-coo** | (17.5) | 0.1 | Tr | Tr | 0 | Tr | Tr | 0 | Tr |
| 81 | **Corn fritters**, *fried in vegetable oil* | 27.8 | 5.0 | 0.4 | 0.3 | 0 | 3.3 | 0.1 | 0.9 | 0.2 |
| 82 | **Corn pudding** | 7.5 | 6.9 | 0.5 | 0.3 | 0 | 3.9 | 0 | 2.2 | 0.3 |
| 83 | **Courgettes with eggs** | 0.1 | 2.0 | 0.8 | 0.8 | 0 | 0.3 | 0 | 0 | 0.2 |
| 84 | **Crumble**, vegetable in milk base | 17.1 | 4.6 | 0.6 | 0.5 | 0 | 2.1 | 0.1 | 1.3 | 0.1 |
| 85 | vegetable in milk base, wholemeal | 14.7 | 4.7 | 0.6 | 0.5 | 0 | 2.2 | Tr | 1.3 | 0.1 |
| 86 | vegetable in tomato base | 14.2 | 4.4 | 1.8 | 1.7 | 0 | 0.8 | Tr | 0.1 | 0.3 |
| 87 | vegetable in tomato base, wholemeal | 11.6 | 4.5 | 1.8 | 1.7 | 0 | 0.9 | Tr | 0.1 | 0.3 |
| 88 | **Curry**, almond | 2.3 | 1.1 | 0.1 | 0.1 | 0 | 0.8 | 0 | 0 | Tr |
| 89 | aubergine | 1.7 | 3.7 | 1.6 | 1.4 | 0 | 0.7 | 0 | 0 | 0.8 |
| 90 | black gram dahl | 6.2 | 0.4 | 0.1 | 0.1 | 0 | 0.2 | 0 | Tr | 0.4 |
| 91 | black gram, whole, Bengali | 6.3 | 1.8 | 0.3 | 0.3 | 0 | 1.2 | 0 | Tr | 0.5 |
| 92 | black gram, whole, Gujerati | 11.7 | 2.3 | 0.8 | 0.8 | 0 | 0.7 | 0 | Tr | 1.0 |

# Vegetable Dishes

**Fibre fractions, phytic acid and fatty acids, g per 100g food**
**Cholesterol, mg per 100g food**

| No. 15- | Food | Dietary fibre, g Southgate method | Dietary fibre, g Englyst method | Fibre fractions, g Cellulose | Non-cellulosic polysaccharide Soluble | Non-cellulosic polysaccharide Insoluble | Lignin | Phytic acid g | Fatty acids, g Satd | Mono- unsatd | Poly- unsatd | Cholesterol mg |
|---|---|---|---|---|---|---|---|---|---|---|---|---|
| 77 | **Coleslaw**, with mayonnaise, retail | 1.6 | 1.4 | 0.5 | 0.6 | 0.3 | 0.2 | Tr | 3.9 | 6.0 | 15.3 | 26 |
| 78 | with reduced calorie dressing, retail | 1.6 | 1.4 | 0.5 | 0.6 | 0.3 | 0.2 | Tr | 0.5 | 1.6 | 2.2 | 0 |
| 79 | with vinaigrette, retail | 1.9 | 1.7 | 0.6 | 0.8 | 0.4 | 0.2 | Tr | 0.5 | 0.9 | 2.3 | 0 |
| 80 | **Coo-coo** | 1.3 | 0.7 | 0.3 | N | N | 0.1 | N | 0.7 | 0.5 | 0.3 | 2 |
| 81 | **Corn fritters**, *fried in vegetable oil* | 2.5 | 1.4 | 0.2 | 0.5 | 0.8 | Tr | 0.05 | 3.1 | 5.2 | 5.9 | 54 |
| 82 | **Corn pudding** | 1.9 | 0.7 | 0.2 | 0.1 | 0.4 | Tr | 0.02 | 1.6 | 1.3 | 0.5 | 64 |
| 83 | **Courgettes with eggs** | (0.9) | 0.9 | 0.3 | 0.4 | 0.2 | 0.1 | 0.04 | 1.6 | 2.8 | 4.9 | 68 |
| 84 | **Crumble**, vegetable in milk base | 2.1 | 1.4 | 0.2 | 0.5 | 0.6 | Tr | 0.05 | 2.9 | 2.7 | 2.8 | 8 |
| 85 | vegetable in milk base, wholemeal | 3.0 | 2.4 | 0.5 | 0.6 | 1.3 | 0.1 | 0.11 | 2.9 | 2.7 | 2.8 | 8 |
| 86 | vegetable in tomato base | (2.1) | 1.9 | 0.5 | 0.9 | 0.5 | Tr | 0.04 | 2.1 | 3.3 | 4.3 | 0 |
| 87 | vegetable in tomato base, wholemeal | (3.0) | 2.9 | 0.7 | 1.0 | 1.2 | 0.1 | 0.11 | 2.1 | 3.3 | 4.4 | 0 |
| 88 | **Curry**, almond | (2.9) | (1.5) | (0.4) | (0.2) | (0.9) | N | (0.23) | 15.8 | 12.1 | 3.6 | 63 |
| 89 | aubergine | 1.7 | 1.5 | 0.5 | 0.8 | 0.2 | 0.1 | 0.01 | 1.1 | 3.5 | 4.8 | 0 |
| 90 | black gram dahl | 1.9 | (1.7) | N | N | N | 0.2 | 0.14 | 2.1 | 0.8 | 0.2 | 8 |
| 91 | black gram, whole, Bengali | 3.2 | (2.6) | N | N | N | 0.7 | 0.12 | 1.7 | 0.6 | 0.2 | 6 |
| 92 | black gram, whole, Gujerati | 6.0 | (4.8) | N | N | N | 1.2 | 0.23 | 4.1 | 1.6 | 0.5 | 16 |

| No. 15- | Food | mg | | | | | | | | | | µg | |
| --- | --- | Na | K | Ca | Mg | P | Fe | Cu | Zn | Cl | Mn | Se | I |
| 77 | **Coleslaw**, with mayonnaise, retail | 160 | 150 | 32 | 3 | 26 | 0.4 | 0.01 | 0.2 | 290 | 0.1 | N | 13 |
| 78 | with reduced calorie dressing, retail | 200 | 160 | 31 | 6 | 19 | 0.3 | 0.01 | 0.1 | 330 | 0.1 | Tr | 2 |
| 79 | with vinaigrette, retail | 130 | 190 | 37 | 7 | 24 | 0.4 | 0.02 | 0.2 | 210 | 0.1 | Tr | 2 |
| 80 | **Coo-coo** | 230 | 56 | 8 | 15 | 31 | 0.4 | 0.04 | 0.3 | 350 | Tr | N | N |
| 81 | **Corn fritters**, *fried in vegetable oil* | 570 | 170 | 76 | 20 | 170 | 1.0 | 0.07 | 0.6 | 750 | 0.2 | 3 | N |
| 82 | **Corn pudding** | 230 | 190 | 63 | 17 | 110 | 0.6 | 0.02 | 0.6 | 340 | 0.1 | 2 | (15) |
| 83 | **Courgettes with eggs** | 94 | 350 | 34 | 22 | 77 | 1.1 | 0.04 | 0.5 | 180 | 0.1 | (3) | (10) |
| 84 | **Crumble**, vegetable in milk base | 210 | 230 | 94 | 18 | 100 | 0.8 | 0.04 | 0.6 | 340 | 0.2 | 1 | (9) |
| 85 | vegetable in milk base, wholemeal | 210 | 260 | 77 | 35 | 140 | 1.1 | 0.09 | 1.0 | 330 | 0.6 | 10 | N |
| 86 | vegetable in tomato base | 210 | 320 | 46 | 17 | 53 | 0.9 | 0.09 | 0.3 | 360 | 0.3 | 1 | (7) |
| 87 | vegetable in tomato base, wholemeal | 210 | 360 | 27 | 35 | 91 | 1.3 | 0.15 | 0.8 | 350 | 0.7 | 10 | N |
| 88 | **Curry**, almond | 260 | 310 | 91 | 73 | 140 | 4.0 | (0.25) | 0.9 | (400) | (0.5) | (1) | N |
| 89 | aubergine | 350 | 270 | 18 | 14 | 29 | 0.5 | 0.06 | 0.2 | 570 | 0.1 | (1) | (4) |
| 90 | black gram dahl | 210 | 140 | 15 | 22 | 44 | 1.1 | 0.12 | 0.5 | 310 | 0.2 | N | N |
| 91 | black gram, whole, Bengali | 240 | 180 | 29 | 29 | 68 | 1.2 | 0.13 | 0.5 | 360 | 0.2 | N | N |
| 92 | black gram, whole, Gujerati | 560 | 380 | 58 | 59 | 130 | 2.4 | 0.28 | 1.0 | 860 | 0.4 | N | N |

| No. 15- | Food | Retinol μg | Carotene μg | Vitamin D μg | Vitamin E mg | Thiamin mg | Ribo-flavin mg | Niacin mg | Trypt 60 mg | Vitamin B6 mg | Vitamin B12 μg | Folate μg | Panto-thenate mg | Biotin μg | Vitamin C mg |
|---|---|---|---|---|---|---|---|---|---|---|---|---|---|---|---|
| 77 | **Coleslaw**, with mayonnaise, retail | 29 | 870 | 0.1 | 6.74 | 0.08 | 0.03 | 0.2 | 0.2 | 0.12 | 0.2 | 21 | N | N | 20 |
| 78 | with reduced calorie dressing, retail | 0 | 755 | 0 | 1.22 | 0.08 | 0.01 | 0.2 | 0.1 | 0.12 | 0 | 20 | 0.14 | 0.1 | 20 |
| 79 | with vinaigrette, retail | 0 | 735 | 0 | 0.86 | 0.10 | 0.01 | 0.3 | 0.2 | 0.14 | 0 | 24 | 0.17 | 0.2 | 24 |
| 80 | **Coo-coo** | 9 | 35 | Tr | 0.02 | 0.06 | 0.02 | 0.2 | 0.3 | 0.01 | Tr | 2 | 0.01 | Tr | 1 |
| 81 | **Corn fritters**, *fried in vegetable oil* | 47 | 53 | 0.2 | 3.13 | 0.09 | 0.10 | 0.9 | 1.3 | 0.09 | 0.4 | 8 | 0.36 | N | Tr |
| 82 | **Corn pudding** | 51 | 59 | 0.3 | 0.44 | 0.04 | 0.14 | 0.6 | 1.1 | 0.09 | 0.6 | 7 | 0.42 | N | 1 |
| 83 | **Courgettes with eggs** | 33 | 540 | 0.3 | 1.56 | 0.11 | 0.08 | 0.3 | 1.0 | 0.13 | 0.4 | 27 | 0.35 | N | 13 |
| 84 | **Crumble**, vegetable in milk base | 67 | 120 | 0.4 | 0.67 | 0.10 | 0.08 | 0.7 | 1.2 | 0.12 | 0.2 | 13 | 0.30 | (1.0) | 7 |
| 85 | vegetable in milk base, wholemeal | 67 | 120 | 0.4 | 0.86 | 0.12 | 0.09 | 1.2 | 1.3 | 0.17 | 0.2 | 16 | 0.37 | (1.8) | 7 |
| 86 | vegetable in tomato base | 65 | 1355 | 0.6 | (1.74) | 0.13 | 0.02 | 0.8 | 0.6 | 0.16 | 0 | 15 | 0.25 | (1.2) | 9 |
| 87 | vegetable in tomato base, wholemeal | 65 | 1355 | 0.6 | (1.94) | 0.15 | 0.03 | 1.4 | 0.7 | 0.21 | 0 | 18 | 0.33 | (2.1) | 9 |
| 88 | **Curry**, almond | 150 | 345 | 0.4 | (5.08) | 0.07 | 0.13 | 0.7 | (0.8) | (0.05) | Tr | (5) | (0.07) | (9.3) | 4 |
| 89 | aubergine | 0 | 215 | 0 | 2.93 | 0.07 | 0.01 | 0.5 | 0.3 | 0.14 | 0 | 9 | 0.14 | N | 4 |
| 90 | black gram dahl | 30 | 75 | Tr | 0.09 | 0.06 | 0.05 | 0.3 | 0.8 | N | Tr | 10 | N | N | Tr |
| 91 | black gram, whole, Bengali | 24 | 110 | Tr | 0.18 | 0.07 | 0.05 | 0.4 | 0.8 | N | Tr | 12 | N | N | 1 |
| 92 | black gram, whole, Gujerati | 60 | 135 | 0.1 | 0.57 | 0.14 | 0.10 | 0.8 | 1.5 | N | Tr | 23 | N | N | 2 |

| No. 15- | Food | Description and main data sources |
|---|---|---|
| 93 | **Curry,** black gram, whole and red kidney bean | Recipe from review of recipe collection |
| 94 | black-eye bean, Gujerati | Recipe from a personal collection |
| 95 | black-eye bean, Punjabi | Recipe from dietary survey records |
| 96 | Bombay potato | Potato, tomato and spices. Recipe from review of recipe collection |
| 97 | cabbage | White cabbage in masala sauce. Recipe from review of recipe collection |
| 98 | cauliflower and potato | Gujerati dish. Cauliflower, potato, tomato and onion. Recipe from a personal collection |
| 99 | chick pea dahl | Punjabi dish. Split chick peas and tomato. Recipe from dietary survey records |
| 100 | chick pea dahl and spinach, *with butter* | Recipe from a personal collection. Split chick peas |
| 101 | *with vegetable oil* | Recipe from a personal collection. Split chick peas |
| 102 | chick pea, whole | Gujerati dish. Recipe from a personal collection |
| 103 | chick pea, whole, basic | Basic UK-type dish. Recipe from review of recipe collection |
| 104 | chick pea, whole and potato | Gujerati dish containing natural yogurt. Recipe from a personal collection |
| 105 | chick pea, whole and tomato, Gujerati, *with butter ghee* | Large white chick peas, tomato and onion. Recipe from a personal collection |
| 106 | *with vegetable oil* | Large white chick peas, tomato and onion. Recipe from a personal collection |
| 107 | chick pea, whole and tomato, Punjabi, *with butter* | Chick peas, onion and tomato. Recipe from dietary survey records |
| 108 | *with vegetable oil* | Chick peas, onion and tomato. Recipe from dietary survey records |
| 109 | courgette and potato | Punjabi dish. Courgette, potato, onion and tomato. Recipe from dietary survey records |

| No. 15- | Food | Water g | Total nitrogen g | Protein g | Fat g | Carbohydrate g | Energy value kcal | kJ |
|---|---|---|---|---|---|---|---|---|
| 93 | **Curry,** black gram, whole and red kidney bean | 76.3 | 0.73 | 4.5 | 5.0 | 8.3 | 92 | 389 |
| 94 | black-eye bean, Gujerati | 67.2 | 1.16 | 7.2 | 4.4 | 16.1 | 127 | 537 |
| 95 | black-eye bean, Punjabi | 72.9 | 0.96 | 6.0 | 4.0 | 13.7 | 111 | 469 |
| 96 | Bombay potato | 73.9 | 0.32 | 2.0 | 6.8 | 13.7 | 117 | 492 |
| 97 | cabbage | 80.9 | 0.31 | 1.9 | 5.0 | 8.1 | 82 | 342 |
| 98 | cauliflower and potato | 83.0 | 0.55 | 3.4 | 2.4 | 6.6 | 59 | 247 |
| 99 | chick pea dahl | 63.1 | 1.19 | 7.4 | 6.1 | 17.7 | 149 | 629 |
| 100 | chick pea dahl and spinach, *with butter* | 63.9 | 0.97 | 6.1 | 11.9 | 12.8 | 179 | 747 |
| 101 | *with vegetable oil* | 62.0 | 0.96 | 6.0 | 14.2 | 12.8 | 199 | 831 |
| 102 | chick pea, whole | 52.7 | 1.54 | 9.6 | 7.5 | 21.3 | 179 | 754 |
| 103 | chick pea, whole, basic | 71.1 | 0.96 | 6.0 | 3.6 | 14.2 | 108 | 455 |
| 104 | chick pea, whole and potato | 72.5 | 0.92 | 5.7 | 5.4 | 12.6 | 119 | 499 |
| 105 | chick pea, whole and tomato, Gujerati, *with butter ghee* | 49.6 | 1.23 | 7.7 | 14.6 | 20.1 | 236 | 986 |
| 106 | *with vegetable oil* | 49.6 | 1.23 | 7.7 | 14.6 | 20.1 | 236 | 987 |
| 107 | chick pea, whole and tomato, Punjabi, *with butter* | 73.6 | 0.90 | 5.6 | 4.3 | 12.4 | 107 | 450 |
| 108 | *with vegetable oil* | 73.1 | 0.89 | 5.6 | 4.9 | 12.4 | 112 | 473 |
| 109 | courgette and potato | 81.8 | 0.30 | 1.9 | 5.2 | 8.7 | 86 | 360 |

# Vegetable Dishes

**Carbohydrate fractions, g per 100g food**

| No. 15- | Food | Starch | Total sugars | Individual sugars | | | | | | Oligo-saccharides |
|---|---|---|---|---|---|---|---|---|---|---|
| | | | | Glucose | Fructose | Galactose | Sucrose | Maltose | Lactose | |
| 93 | **Curry,** black gram, whole and red kidney bean | 6.2 | 1.4 | 0.5 | 0.4 | 0 | 0.5 | 0 | 0 | 0.7 |
| 94 | black-eye bean, Gujerati | 14.6 | 0.9 | 0.1 | Tr | 0 | 0.7 | 0 | 0 | 0.6 |
| 95 | black-eye bean, Punjabi | 12.0 | 1.1 | 0.2 | 0.1 | 0 | 0.7 | 0 | Tr | 0.5 |
| 96 | Bombay potato | 12.4 | 1.2 | 0.5 | 0.4 | 0 | 0.2 | 0 | 0 | Tr |
| 97 | cabbage | 1.5 | 5.9 | 2.6 | 2.3 | 0 | 0.9 | 0 | 0 | 0.7 |
| 98 | cauliflower and potato | 3.8 | 2.6 | 1.3 | 1.1 | 0 | 0.2 | 0 | 0 | 0.2 |
| 99 | chick pea dahl | 14.6 | 1.8 | 0.4 | 0.4 | 0 | 1.0 | 0 | 0 | 1.3 |
| 100 | chick pea dahl and spinach, with butter | 11.2 | 0.9 | 0.1 | 0.1 | 0 | 0.7 | 0 | Tr | 0.8 |
| 101 | with vegetable oil | 11.2 | 0.9 | 0.1 | 0.1 | 0 | 0.7 | 0 | 0 | 0.8 |
| 102 | chick pea, whole | 18.7 | 1.2 | Tr | 0.1 | 0 | 1.0 | 0 | 0 | 1.4 |
| 103 | chick pea, whole, basic | 10.8 | 2.7 | 0.9 | 0.9 | 0 | 0.9 | 0 | 0 | 0.8 |
| 104 | chick pea, whole and potato | 10.2 | 2.1 | Tr | 0.1 | 0.6 | 0.5 | 0 | 0.9 | 0.4 |
| 105 | chick pea, whole and tomato, Gujerati, with butter ghee | 14.7 | 4.0 | 0.8 | 0.9 | 0 | 2.3 | 0 | Tr | 1.4 |
| 106 | with vegetable oil | 14.7 | 4.0 | 0.8 | 0.9 | 0 | 2.3 | 0 | 0 | 1.4 |
| 107 | chick pea, whole and tomato, Punjabi, with butter | 10.8 | 1.1 | 0.2 | 0.2 | 0 | 0.7 | 0 | Tr | 0.5 |
| 108 | with vegetable oil | 10.8 | 1.1 | 0.2 | 0.2 | 0 | 0.7 | 0 | 0 | 0.5 |
| 109 | courgette and potato | 5.9 | 2.4 | 1.0 | 0.9 | 0 | 0.5 | 0 | 0 | 0.4 |

Fibre fractions, phytic acid and fatty acids, g per 100g food
Cholesterol, mg per 100g food

| No. 15- | Food | Dietary fibre, g | | Fibre fractions, g | | | | Phytic acid g | Fatty acids, g | | | Cholesterol mg |
| | | Southgate method | Englyst method | Cellulose | Non-cellulosic polysaccharide | | Lignin | | Satd | Mono-unsatd | Poly-unsatd | |
| | | | | | Soluble | Insoluble | | | | | | |
| 93 | **Curry,** black gram, whole and red kidney bean | 3.4 | (2.7) | N | N | N | N | 0.14 | 0.5 | 1.7 | 2.3 | 0 |
| 94 | black-eye bean, Gujerati | (2.8) | 2.8 | 0.7 | 0.6 | 1.4 | Tr | 0.23 | 0.6 | 1.4 | 2.0 | 0 |
| 95 | black-eye bean, Punjabi | (2.4) | 2.4 | 0.6 | 0.6 | 1.2 | Tr | 0.19 | 2.5 | 0.9 | 0.3 | 9 |
| 96 | Bombay potato | 1.5 | 1.2 | 0.4 | 0.6 | 0.1 | 0.1 | 0.01 | 0.7 | 2.3 | 3.1 | 0 |
| 97 | cabbage | 2.4 | 2.1 | 0.7 | 1.0 | 0.4 | 0.2 | 0.01 | 0.5 | 1.7 | 2.4 | 0 |
| 98 | cauliflower and potato | 1.9 | 1.8 | 0.4 | 0.9 | 0.4 | N | 0.05 | 0.3 | 0.6 | 1.1 | 0 |
| 99 | chick pea dahl | 4.8 | (3.8) | (0.9) | (1.2) | (1.7) | 0.6 | 0.17 | 0.6 | 1.9 | 2.9 | 0 |
| 100 | chick pea dahl and spinach, with butter | (4.3) | (3.4) | (0.9) | (1.1) | (1.4) | 0.5 | 0.12 | 6.9 | 2.8 | 1.1 | 28 |
| 101 | with vegetable oil | (4.3) | (3.4) | (0.9) | (1.1) | (1.4) | 0.5 | 0.12 | 1.5 | 4.8 | 6.9 | 0 |
| 102 | chick pea, whole | 5.8 | 4.5 | 1.1 | 1.4 | 2.1 | 0.8 | 0.23 | 0.7 | 2.3 | 3.6 | 0 |
| 103 | chick pea, whole, basic | 3.9 | 3.3 | 0.9 | 1.1 | 1.3 | 0.5 | 0.11 | 0.4 | 1.0 | 1.7 | 0 |
| 104 | chick pea, whole and potato | 2.9 | 2.4 | 0.6 | 0.8 | 1.0 | 0.4 | 0.10 | 0.8 | 1.7 | 2.4 | 2 |
| 105 | chick pea, whole and tomato, Gujerati, *with butter ghee* | 5.2 | 4.1 | 1.0 | 1.4 | 1.7 | 0.7 | 0.18 | 8.5 | 3.5 | 1.4 | 35 |
| 106 | *with vegetable oil* | 5.2 | 4.1 | 1.0 | 1.4 | 1.7 | 0.7 | 0.18 | 1.5 | 4.9 | 7.1 | 0 |
| 107 | chick pea, whole and tomato, Punjabi, *with butter* | 3.5 | 2.9 | 0.7 | 1.0 | 1.2 | 0.5 | 0.10 | 2.0 | 1.0 | 0.8 | 8 |
| 108 | *with vegetable oil* | 3.5 | 2.9 | 0.7 | 1.0 | 1.2 | 0.5 | 0.10 | 0.5 | 1.5 | 2.4 | 0 |
| 109 | courgette and potato | (1.4) | 1.2 | 0.4 | 0.6 | 0.2 | 0.1 | 0.03 | 0.6 | 1.7 | 2.5 | 0 |

| No. 15- | Food | Na | K | Ca | Mg | P | Fe | Cu | Zn | Cl | Mn | Se | I |
|---|---|---|---|---|---|---|---|---|---|---|---|---|---|
| | | | | | | mg | | | | | | µg | |
| 93 | **Curry**, black gram, whole and red kidney bean | 310 | 240 | 35 | 33 | 75 | 1.6 | 0.15 | 0.6 | 470 | 0.3 | (1) | (1) |
| 94 | black-eye bean, Gujerati | 300 | 290 | 19 | 40 | 120 | 1.8 | 0.18 | 0.9 | 450 | 0.4 | 2 | N |
| 95 | black-eye bean, Punjabi | N | 240 | 16 | 30 | 97 | (1.3) | 0.15 | 0.8 | N | 0.3 | 2 | N |
| 96 | Bombay potato | 420 | 360 | 14 | 21 | 39 | 1.0 | 0.08 | 0.3 | 690 | 0.2 | 1 | 4 |
| 97 | cabbage | 310 | 330 | 45 | 13 | 39 | 0.7 | 0.05 | 0.3 | 510 | 0.2 | Tr | (4) |
| 98 | cauliflower and potato | 210 | 410 | 21 | 20 | 62 | 0.8 | 0.06 | 0.6 | 360 | 0.3 | Tr | (1) |
| 99 | chick pea dahl | 19 | 400 | 59 | 46 | 110 | 2.1 | 0.33 | 1.1 | 37 | 0.8 | 1 | N |
| 100 | chick pea dahl and spinach, *with butter* | 140 | 330 | 87 | 43 | 90 | 2.3 | 0.25 | 0.9 | 180 | 0.8 | (1) | N |
| 101 | *with vegetable oil* | 43 | 320 | 85 | 43 | 87 | 2.3 | 0.25 | 0.9 | 33 | 0.8 | (1) | N |
| 102 | chick pea, whole | 390 | 490 | 89 | 67 | 150 | 4.0 | 0.44 | 1.4 | 600 | 1.1 | 1 | N |
| 103 | chick pea, whole, basic | 12 | 300 | 41 | 29 | 69 | 1.9 | 0.21 | 0.9 | 34 | 0.6 | 1 | N |
| 104 | chick pea, whole and potato | 19 | 240 | 65 | 26 | 85 | 1.4 | 0.16 | 0.8 | 40 | 0.4 | 1 | N |
| 105 | chick pea, whole and tomato, *Gujerati, with butter ghee* | 590 | 470 | 63 | 53 | 120 | 2.3 | 0.34 | 1.1 | 930 | 0.9 | 1 | N |
| 106 | *with vegetable oil* | 590 | 470 | 63 | 53 | 120 | 2.3 | 0.34 | 1.1 | 930 | 0.9 | 1 | N |
| 107 | chick pea, whole and tomato, *Punjabi, with butter* | N | 210 | 32 | 25 | 58 | (1.4) | 0.19 | 0.8 | N | 0.5 | 1 | N |
| 108 | *with vegetable oil* | N | 210 | 32 | 25 | 57 | (1.4) | 0.19 | 0.8 | N | 0.5 | 1 | N |
| 109 | courgette and potato | N | 350 | 17 | 17 | 42 | (0.6) | 0.05 | 0.3 | N | 0.1 | (1) | N |

| No. 15- | Food | Retinol µg | Carotene µg | Vitamin D µg | Vitamin E mg | Thiamin mg | Ribo-flavin mg | Niacin mg | Trypt 60 mg | Vitamin B6 mg | Vitamin B12 µg | Folate µg | Panto-thenate mg | Biotin µg | Vitamin C mg |
|---|---|---|---|---|---|---|---|---|---|---|---|---|---|---|---|
| 93 | **Curry**, black gram, whole and red kidney bean | 0 | 145 | 0 | (1.30) | 0.09 | 0.05 | 0.5 | 0.8 | (0.04) | 0 | 13 | (0.06) | (0.2) | 2 |
| 94 | black-eye bean, Gujerati | 0 | 105 | 0 | (0.89) | 0.12 | 0.04 | 0.4 | 1.6 | 0.07 | 0 | 84 | 0.19 | 4.5 | Tr |
| 95 | black-eye bean, Punjabi | 35 | 64 | Tr | (0.16) | 0.14 | 0.03 | 0.4 | 1.3 | 0.08 | Tr | 140 | 0.22 | 4.7 | 1 |
| 96 | Bombay potato | 0 | 225 | 0 | 1.84 | 0.15 | 0.02 | 0.5 | 0.4 | 0.29 | 0 | 15 | 0.27 | 0.5 | 6 |
| 97 | cabbage | 0 | 195 | 0 | 1.75 | 0.12 | 0.01 | 0.6 | 0.3 | 0.20 | 0 | 16 | 0.21 | 0.7 | 15 |
| 98 | cauliflower and potato | 0 | 98 | 0 | 0.69 | 0.15 | 0.04 | 0.6 | 0.8 | 0.26 | 0 | 30 | 0.45 | 1.1 | 19 |
| 99 | chick pea dahl | 0 | 210 | 0 | 2.16 | 0.13 | 0.07 | 0.7 | 1.0 | 0.17 | 0 | 32 | 0.46 | N | 2 |
| 100 | chick pea dahl and spinach, with butter | 100 | 1030 | 0.1 | 1.40 | 0.09 | 0.06 | 0.6 | 0.9 | 0.12 | Tr | 33 | (0.36) | N | 2 |
| 101 | with vegetable oil | 0 | 975 | 0 | 4.17 | 0.09 | 0.06 | 0.6 | 0.9 | 0.12 | 0 | 33 | (0.36) | N | 2 |
| 102 | chick pea, whole | 0 | 160 | 0 | 2.36 | 0.15 | 0.09 | 0.8 | 1.3 | 0.18 | 0 | 38 | 0.54 | N | 1 |
| 103 | chick pea, whole, basic | 0 | 235 | 0 | 1.65 | 0.09 | 0.04 | 0.7 | 0.8 | 0.14 | 0 | 21 | 0.23 | N | 3 |
| 104 | chick pea, whole and potato | 5 | 18 | Tr | 1.50 | 0.07 | 0.07 | 0.4 | 0.9 | 0.10 | Tr | 17 | 0.24 | N | 1 |
| 105 | chick pea, whole and tomato, Gujerati, *with butter ghee* | 84 | 325 | 0.2 | 1.87 | 0.15 | 0.07 | 0.9 | 1.1 | 0.21 | Tr | 34 | 0.51 | N | 4 |
| 106 | *with vegetable oil* | 0 | 260 | 0 | 4.48 | 0.15 | 0.07 | 0.9 | 1.1 | 0.21 | 0 | 34 | 0.51 | N | 4 |
| 107 | chick pea, whole and tomato, Punjabi, *with butter* | 28 | 80 | Tr | 0.89 | 0.08 | 0.05 | 0.6 | 0.7 | 0.11 | Tr | 36 | 0.21 | N | 1 |
| 108 | *with vegetable oil* | 0 | 65 | 0 | 1.66 | 0.08 | 0.05 | 0.6 | 0.7 | 0.11 | 0 | 36 | 0.21 | N | 1 |
| 109 | courgette and potato | 0 | 365 | 0 | N | 0.13 | 0.01 | 0.5 | 0.4 | 0.22 | 0 | 19 | 0.18 | N | 8 |

| No. | Food | Description and main data sources |
| --- | --- | --- |
| **15-** | | |
| 110 | **Curry**, dudhi, kofta | Gujerati dish. Fried dudhi balls in a tomato and onion sauce. Recipe from a personal collection |
| 111 | egg, *with butter* | Punjabi dish. Egg, onion and tomato. Recipe from review of recipe collection |
| 112 | *with vegetable oil* | Punjabi dish. Egg, onion and tomato. Recipe from review of recipe collection |
| 113 | egg, in sweet sauce | Eggs in basic UK-type curry sauce. Recipe from review of recipe collection |
| 114 | egg and potato | Egg, potato and tomato. Recipe from review of recipe collection |
| 115 | gobi aloo sag, retail | Cauliflower, onion, potato, spinach and tomato. Recipe from manufacturer |
| 116 | green bean | Gujerati dish. Frozen green beans in masala sauce. Recipe from review of recipe collection |
| 117 | karela | Punjabi dish. Karela, onion and tomato. Recipe from dietary survey records |
| 118 | lentil, red/masoor dahl, *with butter* | Bangladeshi dish. Recipe adapted from Carlson and de Wet, 1991. Thin-medium consistency |
| 119 | *with vegetable oil* | Bangladeshi dish. Recipe adapted from Carlson and de Wet 1991. Thin-medium consistency |
| 120 | lentil, red/masoor dahl, Punjabi | Recipe from review of recipe collection. Thick consistency |
| 121 | lentil, red/masoor dahl and tomato, *with butter* | Recipe from review of recipe collection. Thin consistency |
| 122 | *with vegetable oil* | Recipe from review of recipe collection. Thin consistency |
| 123 | lentil, red/masoor dahl and tomato, Punjabi | Recipe from review of recipe collection. Thick consistency |
| 124 | lentil, red/masoor dahl and mung bean dahl | Recipe from review of recipe collection. Thin consistency |

**Vegetable Dishes**

**Composition of food per 100g**

| No. Food 15- | Water g | Total nitrogen g | Protein g | Fat g | Carbohydrate g | Energy value kcal | Energy value kJ |
|---|---|---|---|---|---|---|---|
| 110 **Curry**, dudhi, kofta | 76.0 | 0.42 | 2.6 | 7.5 | 9.4 | 113 | 469 |
| 111 egg, *with butter* | 76.4 | 0.79 | 4.9 | 11.0 | 4.7 | 134 | 557 |
| 112 *with vegetable oil* | 75.0 | 0.78 | 4.8 | 12.6 | 4.7 | 149 | 617 |
| 113 egg, in sweet sauce | 78.6 | 0.92 | 5.7 | 7.7 | 5.8 | 113 | 473 |
| 114 egg and potato | 73.4 | 1.09 | 6.8 | 9.9 | 6.3 | 136 | 568 |
| 115 gobi aloo sag, retail | 79.4 | 0.35 | 2.2 | 6.9 | 7.1 | 95 | 397 |
| 116 green bean | 77.7 | 0.27 | 1.7 | 12.7 | 3.6 | 131 | 540 |
| 117 karela | 79.8 | 0.28 | 1.7 | 9.8 | 4.6 | 111 | 462 |
| 118 lentil, red/masoor dahl, *with butter* | 62.6 | 1.22 | 7.6 | 6.5 | 19.2 | 160 | 673 |
| 119 *with vegetable oil* | 61.4 | 1.21 | 7.6 | 7.9 | 19.2 | 172 | 722 |
| 120 lentil, red/masoor dahl, Punjabi | 63.7 | 1.15 | 7.2 | 4.6 | 19.2 | 139 | 587 |
| 121 lentil, red/masoor dahl and tomato, *with butter* | 78.9 | 0.64 | 4.0 | 4.9 | 9.7 | 94 | 397 |
| 122 *with vegetable oil* | 78.1 | 0.63 | 4.0 | 5.9 | 9.7 | 103 | 434 |
| 123 lentil, red/masoor dahl and tomato, Punjabi | 78.8 | 0.52 | 3.2 | 5.5 | 8.4 | 92 | 388 |
| 124 lentil, red/masoor dahl and mung bean dahl | 75.2 | 0.76 | 4.8 | 6.7 | 9.9 | 114 | 478 |

| No. 15- | Food | Starch | Total sugars | Individual sugars | | | | | | Oligo-saccharides |
|---|---|---|---|---|---|---|---|---|---|---|
| | | | | Glucose | Fructose | Galactose | Sucrose | Maltose | Lactose | |
| 110 | **Curry**, dudhi, kofta | N | N | N | N | 0 | N | 0 | 0 | 1.2 |
| 111 | egg, *with butter* | 0.3 | 3.3 | 1.3 | 1.1 | 0 | 0.9 | 0 | Tr | 1.1 |
| 112 | *with vegetable oil* | 0.3 | 3.3 | 1.3 | 1.1 | 0 | 0.9 | 0 | 0 | 1.1 |
| 113 | egg, in sweet sauce | 2.2 | 3.3 | 1.3 | 1.5 | 0 | 0.5 | 0 | 0 | 0.3 |
| 114 | egg and potato | 4.6 | 1.5 | 0.6 | 0.5 | 0 | 0.4 | 0 | 0 | 0.3 |
| 115 | gobi aloo sag, retail | 3.3 | 3.3 | 1.1 | 1.0 | 0 | 1.2 | 0 | 0 | 0.5 |
| 116 | green bean | 0.6 | 3.0 | 0.7 | 0.9 | 0 | 1.3 | 0 | 0 | Tr |
| 117 | karela | 0.5 | 3.1 | 1.2 | 1.0 | 0 | 0.8 | 0 | 0 | 1.0 |
| 118 | lentil, red/masoor dahl, *with butter* | 15.5 | 2.2 | 0.6 | 0.5 | 0 | 1.1 | 0 | Tr | 1.5 |
| 119 | *with vegetable oil* | 15.5 | 2.2 | 0.6 | 0.5 | 0 | 1.1 | 0 | 0 | 1.5 |
| 120 | lentil, red/masoor dahl, Punjabi | 13.7 | 3.5 | 1.1 | 0.9 | 0 | 1.6 | 0 | 0 | 2.0 |
| 121 | lentil, red/masoor dahl and tomato, *with butter* | 8.0 | 1.0 | 0.3 | 0.2 | 0 | 0.5 | 0 | Tr | 0.7 |
| 122 | *with vegetable oil* | 8.0 | 1.0 | 0.3 | 0.2 | 0 | 0.5 | 0 | 0 | 0.7 |
| 123 | lentil, red/masoor dahl and tomato, Punjabi | 5.7 | 2.4 | 0.9 | 1.0 | 0 | 0.4 | 0 | 0 | 0.4 |
| 124 | lentil, red/masoor dahl and mung bean dahl | (8.2) | (0.9) | (0.2) | (0.2) | 0 | (0.5) | 0 | 0 | 0.8 |

**Fibre fractions, phytic acid and fatty acids, g per 100g food**
**Cholesterol, mg per 100g food**

| No. 15- | Food | Dietary fibre, g | | Fibre fractions, g | | | | Phytic acid g | Fatty acids, g | | | Cholesterol mg |
|---|---|---|---|---|---|---|---|---|---|---|---|---|
| | | Southgate method | Englyst method | Cellulose | Non-cellulosic polysaccharide Soluble | Non-cellulosic polysaccharide Insoluble | Lignin | | Satd | Mono-unsatd | Poly-unsatd | |
| 110 | **Curry**, dudhi, kofta | (3.1) | (2.8) | N | N | N | 0.2 | 0.08 | 2.2 | 2.2 | 2.4 | 7 |
| 111 | egg, *with butter* | 0.9 | 0.8 | 0.3 | 0.4 | 0.1 | Tr | 0.02 | 5.8 | 3.3 | 0.7 | 140 |
| 112 | *with vegetable oil* | 0.9 | 0.8 | 0.3 | 0.4 | 0.1 | Tr | 0.02 | 1.9 | 4.7 | 4.8 | 120 |
| 113 | egg, in sweet sauce | 1.1 | 0.9 | 0.2 | 0.3 | 0.4 | Tr | 0.01 | 2.2 | 2.8 | 1.7 | 153 |
| 114 | egg and potato | 0.9 | 0.7 | 0.2 | 0.4 | 0.1 | Tr | 0.01 | 1.9 | 3.9 | 2.9 | 175 |
| 115 | gobi aloo sag, retail | (1.8) | 1.4 | 0.5 | 0.7 | 0.3 | 0.1 | 0.03 | 0.7 | 2.4 | 3.3 | 0 |
| 116 | green bean | 2.2 | 1.6 | 0.6 | 0.7 | 0.4 | N | 0.01 | 1.3 | 4.4 | 5.9 | 0 |
| 117 | karela | 3.3 | 2.5 | 0.9 | 1.2 | 0.5 | 0.3 | 0.02 | 1.2 | 2.4 | 5.7 | 0 |
| 118 | lentil, red/masoor dahl, *with butter* | 3.5 | 1.8 | 0.4 | 0.6 | 0.9 | 0.2 | 0.09 | 4.0 | 1.5 | 0.4 | 16 |
| 119 | *with vegetable oil* | 3.5 | 1.8 | 0.4 | 0.6 | 0.9 | 0.2 | 0.09 | 0.8 | 2.7 | 3.7 | 0 |
| 120 | lentil, red/masoor dahl, Punjabi | 3.6 | 2.0 | 0.5 | 0.8 | 0.8 | 0.1 | 0.09 | 0.5 | 1.5 | 2.1 | 0 |
| 121 | lentil, red/masoor dahl and tomato, *with butter* | 1.8 | 0.9 | 0.2 | 0.3 | 0.5 | 0.1 | 0.05 | 3.1 | 1.2 | 0.3 | 12 |
| 122 | *with vegetable oil* | 1.8 | 0.9 | 0.2 | 0.3 | 0.5 | 0.1 | 0.05 | 0.6 | 2.0 | 2.8 | 0 |
| 123 | lentil, red/masoor dahl and tomato, Punjabi | 1.9 | 1.2 | 0.3 | 0.5 | 0.4 | N | 0.04 | 0.6 | 1.9 | 2.6 | 0 |
| 124 | lentil, red/masoor dahl and mung bean dahl | 2.2 | (1.6) | N | N | N | N | 0.10 | 0.7 | 2.3 | 3.2 | 0 |

| No. 15- | Food | Na | K | Ca | Mg | P | Fe | Cu | Zn | Cl | Mn | Se | I |
|---|---|---|---|---|---|---|---|---|---|---|---|---|---|
| | | | | | | mg | | | | | | µg | |
| 110 | **Curry**, dudhi, kotta | 510 | 280 | 37 | 23 | 53 | 1.3 | 0.11 | 0.7 | 810 | N | N | N |
| 111 | egg, *with butter* | 280 | 200 | 37 | 12 | 87 | 1.1 | 0.07 | 0.6 | 420 | 0.1 | 4 | (22) |
| 112 | *with vegetable oil* | 210 | 200 | 35 | 12 | 84 | 1.1 | 0.07 | 0.6 | 320 | 0.1 | 4 | (20) |
| 113 | egg, in sweet sauce | 65 | 150 | 39 | 11 | 95 | 1.7 | 0.08 | 0.6 | 83 | 0.1 | 5 | 22 |
| 114 | egg and potato | 370 | 250 | 40 | 18 | 120 | 1.8 | 0.08 | 0.8 | 560 | 0.1 | 5 | 27 |
| 115 | gobi aloo sag, retail | 530 | 310 | 36 | 21 | 42 | 0.7 | 0.05 | 0.4 | 820 | 0.2 | Tr | 3 |
| 116 | green bean | 200 | 210 | 35 | 18 | 35 | 1.5 | 0.02 | 0.2 | 320 | N | N | N |
| 117 | karela | N | 340 | 24 | 23 | 49 | (1.2) | 0.20 | 0.4 | N | 0.3 | N | N |
| 118 | lentil, red/masoor dahl, *with butter* | 330 | 280 | 24 | 30 | 110 | 2.6 | 0.19 | 1.0 | 510 | N | 2 | N |
| 119 | *with vegetable oil* | 270 | 280 | 23 | 30 | 110 | 2.6 | 0.19 | 1.0 | 430 | N | 2 | N |
| 120 | lentil, red/masoor dahl, Punjabi | 720 | 290 | 29 | 31 | 110 | 2.4 | 0.19 | 1.0 | 1130 | N | (2) | N |
| 121 | lentil, red/masoor dahl and tomato, *with butter* | 230 | 150 | 14 | 16 | 57 | 1.5 | 0.10 | 0.5 | 360 | N | (1) | N |
| 122 | *with vegetable oil* | 190 | 150 | 13 | 16 | 56 | 1.5 | 0.10 | 0.5 | 290 | N | (1) | N |
| 123 | lentil, red/masoor dahl and tomato, Punjabi | 210 | 240 | 17 | 16 | 54 | 1.5 | 0.09 | 0.5 | 340 | N | 1 | N |
| 124 | lentil, red/masoor dahl and mung bean dahl | 210 | 210 | 15 | 23 | 70 | 1.6 | 0.12 | 0.6 | 330 | N | (2) | N |

| No. 15- | Food | Retinol µg | Carotene µg | Vitamin D µg | Vitamin E mg | Thiamin mg | Riboflavin mg | Niacin mg | Trypt 60 mg | Vitamin B6 mg | Vitamin B12 µg | Folate µg | Pantothenate mg | Biotin µg | Vitamin C mg |
|---|---|---|---|---|---|---|---|---|---|---|---|---|---|---|---|
| 110 | **Curry**, dudhi, kofta | 17 | 88 | Tr | 1.79 | 0.09 | 0.03 | 0.7 | 0.4 | 0.13 | 0 | 12 | (0.29) | N | 5 |
| 111 | egg, *with butter* | 130 | 200 | 0.6 | 0.90 | 0.08 | 0.10 | 0.5 | 1.3 | 0.13 | 0.3 | 11 | 0.41 | 4.6 | 2 |
| 112 | *with vegetable oil* | 59 | 165 | 0.6 | 2.88 | 0.08 | 0.10 | 0.5 | 1.3 | 0.13 | 0.3 | 11 | 0.40 | 4.6 | 2 |
| 113 | egg, in sweet sauce | 75 | 19 | 0.7 | 1.19 | 0.05 | 0.15 | 0.2 | 1.6 | 0.08 | 0.4 | 17 | 0.55 | 6.7 | 1 |
| 114 | egg and potato | 86 | 160 | 0.8 | 1.87 | 0.10 | 0.14 | 0.4 | 1.9 | 0.18 | 0.5 | 16 | 0.60 | 6.2 | 3 |
| 115 | gobi aloo sag, retail | 0 | 505 | 0 | 2.04 | 0.11 | 0.02 | 0.6 | 0.5 | 0.19 | 0 | 23 | (0.26) | (0.7) | 10 |
| 116 | green bean | 0 | (360) | 0 | 3.16 | 0.04 | 0.04 | 0.6 | 0.4 | 0.04 | 0 | (28) | 0.07 | 0.7 | 5 |
| 117 | karela | 0 | 385 | 0 | N | 0.11 | 0.03 | 0.6 | 0.3 | N | 0 | 20 | N | N | 64 |
| 118 | lentil, red/masoor dahl, *with butter* | 59 | 95 | 0.1 | N | 0.15 | 0.05 | 0.7 | 1.1 | 0.19 | Tr | (7) | 0.36 | N | 1 |
| 119 | *with vegetable oil* | 0 | 64 | 0 | N | 0.15 | 0.05 | 0.7 | 1.1 | 0.19 | 0 | (7) | 0.35 | N | 1 |
| 120 | lentil, red/masoor dahl, Punjabi | 0 | 250 | 0 | 1.11 | 0.16 | 0.05 | 0.8 | 1.0 | 0.21 | 0 | 9 | 0.34 | N | 1 |
| 121 | lentil, red/masoor dahl and tomato, *with butter* | 45 | 97 | Tr | 0.20 | 0.08 | 0.03 | 0.4 | 0.6 | 0.09 | Tr | 3 | 0.19 | N | 1 |
| 122 | *with vegetable oil* | 0 | 73 | 0 | 1.43 | 0.08 | 0.03 | 0.4 | 0.5 | 0.09 | 0 | 3 | 0.19 | N | 1 |
| 123 | lentil, red/masoor dahl and tomato, Punjabi | 0 | 395 | 0 | 1.86 | 0.08 | 0.02 | 0.6 | 0.4 | 0.11 | 0 | 5 | 0.19 | N | 5 |
| 124 | lentil, red/masoor dahl and mung bean dahl | 0 | 76 | 0 | 1.55 | (0.07) | (0.04) | (0.4) | 0.7 | (0.09) | 0 | (8) | (0.24) | N | Tr |

| No. 15- | Food | Description and main data sources |
|---------|------|-----------------------------------|
| 125 | **Curry,** lentil, red/masoor dahl, mung bean dahl and tomato | Punjabi dish. Recipe from dietary survey records |
| 126 | lentil, red/masoor dahl and vegetable | Basic UK-type dish. Lentil, tomato, carrot and pepper. Recipe from review of recipe collection |
| 127 | lentil, whole/masoor, Gujerati | Recipe from a personal collection. Thick consistency |
| 128 | lentil, whole/masoor, Punjabi | Recipe from review of recipe collection. Medium consistency |
| 129 | mung bean dahl, Bengali | Recipe from Tan et al, 1985. Split mung beans. Thin consistency |
| 130 | mung bean dahl, Punjabi | Recipe from dietary survey records. Thick consistency |
| 131 | mung bean dahl and spinach | Gujerati dish. Recipe from a personal collection. Split mung beans |
| 132 | mung bean dahl and tomato | Punjabi dish. Recipe from dietary survey records. Split mung beans |
| 133 | mung bean, whole, Gujerati | Recipe from a personal collection. Thick consistency |
| 134 | mung bean, whole, Punjabi | Recipe from Tan et al, 1985. Medium consistency |
| 135 | mung bean, whole and turnip leaves | Recipe from Wharton et al, 1983 |
| 136 | okra | Okra in masala sauce. Recipe from review of recipe collection |
| 137 | pea and potato | Recipe from Wharton et al, 1983. Canned peas |
| 138 | pigeon pea dahl, *with butter* | Recipe from Tan et al, 1985. Split pigeon peas |
| 139 | *with vegetable oil* | Recipe from Tan et al, 1985. Split pigeon peas |
| 140 | pigeon pea dahl and tomato | Punjabi dish. Recipe from review of recipe collection. Thin consistency. Split pigeon peas |

| No. 15- | Food | Water g | Total nitrogen g | Protein g | Fat g | Carbohydrate g | Energy value kcal | kJ |
|---|---|---|---|---|---|---|---|---|
| 125 | **Curry**, lentil, red/masoor dahl, mung bean dahl and tomato | 78.9 | 0.71 | 4.4 | 4.3 | (9.7) | 92 | 387 |
| 126 | lentil, red/masoor dahl and vegetable | 72.3 | 0.92 | 5.8 | 3.8 | 14.7 | 110 | 466 |
| 127 | lentil, whole/masoor, Gujerati | 66.9 | 1.06 | 6.6 | 5.4 | 15.1 | 130 | 550 |
| 128 | lentil, whole/masoor, Punjabi | 68.4 | 0.92 | 5.7 | 8.4 | 11.8 | 139 | 583 |
| 129 | mung bean dahl, Bengali | 81.9 | 0.67 | 4.2 | 3.3 | (7.4) | 73 | 310 |
| 130 | mung bean dahl, Punjabi | 73.3 | 1.00 | 6.3 | 3.1 | (12.4) | 99 | 421 |
| 131 | mung bean dahl and spinach | 68.0 | 1.24 | 7.7 | 2.9 | (12.4) | 101 | 427 |
| 132 | mung bean dahl and tomato | 74.8 | 0.90 | 5.6 | 3.6 | (11.2) | 95 | 403 |
| 133 | mung bean, whole, Gujerati | 68.7 | 0.95 | 6.0 | 5.9 | 13.2 | 125 | 525 |
| 134 | mung bean, whole, Punjabi | 80.1 | 0.67 | 4.2 | 2.8 | 8.8 | 74 | 312 |
| 135 | mung bean, whole and turnip leaves | 68.3 | 0.69 | 4.3 | 14.2 | 5.7 | 163 | 679 |
| 136 | okra | 76.8 | 0.38 | 2.6 | 8.9 | 5.3 | 109 | 452 |
| 137 | pea and potato | 73.0 | 0.40 | 2.5 | 11.1 | 10.1 | 146 | 608 |
| 138 | pigeon pea dahl, *with butter* | 74.2 | 0.73 | 4.6 | 4.6 | 13.8 | 109 | 460 |
| 139 | *with vegetable oil* | 73.4 | 0.73 | 4.6 | 5.5 | 13.8 | 117 | 493 |
| 140 | pigeon pea dahl and tomato | 86.1 | 0.38 | 2.4 | 2.2 | 7.6 | 57 | 243 |

**Carbohydrate fractions, g per 100g food**

| No. 15- | Food | Starch | Total sugars | Individual sugars | | | | | | Oligo-saccharides |
|---|---|---|---|---|---|---|---|---|---|---|
| | | | | Glucose | Fructose | Galactose | Sucrose | Maltose | Lactose | |
| 125 | **Curry**, lentil, red/masoor dahl, mung bean dahl and tomato | (8.5) | (0.8) | (0.2) | (0.2) | 0 | (0.3) | 0 | 0 | 0.4 |
| 126 | lentil, red/masoor dahl and vegetable | 11.4 | 2.4 | 0.7 | 0.8 | 0 | 0.9 | 0 | 0 | 0.9 |
| 127 | lentil, whole/masoor, Gujerati | 11.1 | 2.9 | 0.7 | 0.6 | 0 | 1.6 | 0 | Tr | 1.1 |
| 128 | lentil, whole/masoor, Punjabi | 9.7 | 1.2 | 0.4 | 0.4 | 0 | 0.5 | 0 | 0 | 0.9 |
| 129 | mung bean dahl, Bengali | (6.2) | (0.5) | (0.1) | (0.1) | 0 | (0.2) | 0 | Tr | 0.7 |
| 130 | mung bean dahl, Punjabi | (11.2) | (0.6) | (0.2) | (0.1) | 0 | (0.3) | 0 | 0 | 0.6 |
| 131 | mung bean dahl and spinach | (8.4) | (3.3) | (0.4) | (0.5) | 0 | (2.4) | 0 | 0 | 0.8 |
| 132 | mung bean dahl and tomato | (9.8) | (0.8) | (0.3) | (0.3) | 0 | (0.3) | 0 | 0 | 0.6 |
| 133 | mung bean, whole, Gujerati | 9.5 | 2.5 | 0.7 | 0.6 | 0 | 1.2 | 0 | 0 | 1.2 |
| 134 | mung bean, whole, Punjabi | 6.9 | 1.0 | 0.3 | 0.3 | 0 | 0.3 | 0 | Tr | 0.9 |
| 135 | mung bean, whole and turnip leaves | 5.0 | 0.3 | 0.1 | 0.1 | 0 | 0.1 | 0 | Tr | 0.2 |
| 136 | okra | 0.6 | 4.1 | 1.4 | 1.5 | 0 | 1.1 | 0 | 0 | 0.7 |
| 137 | pea and potato | 6.2 | 2.6 | 0.6 | 0.5 | 0 | 1.5 | 0 | 0 | 1.3 |
| 138 | pigeon pea dahl, *with butter* | 10.8 | 2.4 | 0.1 | 0.1 | 0 | 2.1 | 0 | Tr | 0.5 |
| 139 | *with vegetable oil* | 10.8 | 2.4 | 0.1 | 0.1 | 0 | 2.1 | 0 | 0 | 0.5 |
| 140 | pigeon pea dahl and tomato | 5.6 | 1.8 | 0.1 | 0.1 | 0 | 1.5 | 0 | 0 | 0.2 |

**Fibre fractions, phytic acid and fatty acids, g per 100g food**
**Cholesterol, mg per 100g food**

| No. 15- | Food | Dietary fibre, g Southgate method | Englyst method | Fibre fractions, g Cellulose | Non-cellulosic polysaccharide Soluble | Insoluble | Lignin | Phytic acid g | Fatty acids, g Satd | Mono-unsatd | Poly-unsatd | Cholesterol mg |
|---|---|---|---|---|---|---|---|---|---|---|---|---|
| 125 | **Curry**, lentil, red/masoor dahl, mung bean dahl and tomato | 2.1 | (1.6) | N | N | N | N | 0.06 | 0.5 | 1.5 | 2.1 | 0 |
| 126 | lentil, red/masoor dahl and vegetable | (3.1) | 1.8 | 0.5 | 0.6 | 0.7 | 0.2 | 0.07 | 0.4 | 1.3 | 1.8 | 0 |
| 127 | lentil, whole/masoor, Gujerati | (2.6) | 2.6 | 0.9 | 0.6 | 1.0 | N | 0.13 | 3.3 | 1.3 | 0.4 | 13 |
| 128 | lentil, whole/masoor, Punjabi | (2.2) | 2.2 | 0.7 | 0.5 | 0.9 | N | 0.11 | 0.9 | 2.9 | 3.9 | 0 |
| 129 | mung bean dahl, Bengali | 1.9 | (1.7) | N | N | N | 0.2 | 0.12 | 2.1 | 0.8 | 0.2 | 8 |
| 130 | mung bean dahl, Punjabi | 3.1 | (2.9) | N | N | N | 0.2 | 0.11 | 0.4 | 1.0 | 1.5 | 0 |
| 131 | mung bean dahl and spinach | 5.4 | (3.8) | N | N | N | 0.3 | 0.16 | 0.3 | 0.8 | 1.4 | 0 |
| 132 | mung bean dahl and tomato | 2.9 | (2.6) | N | N | N | 0.2 | 0.10 | 0.4 | 1.2 | 1.7 | 0 |
| 133 | mung bean, whole, Gujerati | 3.5 | 2.6 | 1.0 | 0.8 | 0.8 | 0.5 | 0.17 | 0.7 | 2.0 | 2.8 | 0 |
| 134 | mung bean, whole, Punjabi | 2.6 | 1.9 | 0.7 | 0.6 | 0.6 | 0.4 | 0.12 | 1.7 | 0.6 | 0.2 | 6 |
| 135 | mung bean, whole and turnip leaves | 3.4 | N | N | N | N | N | 0.07 | 4.7 | 4.3 | 4.1 | 16 |
| 136 | okra | 3.6 | 3.2 | 1.0 | 1.9 | 0.3 | 0.3 | 0.01 | 1.0 | 2.9 | 4.1 | 0 |
| 137 | pea and potato | 2.5 | 2.2 | 1.1 | 0.8 | 0.3 | Tr | 0.02 | 1.2 | 3.8 | 5.3 | 0 |
| 138 | pigeon pea dahl, *with butter* | 2.3 | (2.1) | N | N | N | 0.2 | 0.20 | 2.7 | 1.1 | 0.3 | 11 |
| 139 | *with vegetable oil* | 2.3 | (2.1) | N | N | N | 0.2 | 0.20 | 0.6 | 1.8 | 2.6 | 0 |
| 140 | pigeon pea dahl and tomato | 1.2 | (1.1) | N | N | N | 0.1 | 0.10 | 0.3 | 0.7 | 1.1 | 0 |

| No. 15- | Food | mg | | | | | | | | | | µg | |
|---|---|---|---|---|---|---|---|---|---|---|---|---|---|
| | | Na | K | Ca | Mg | P | Fe | Cu | Zn | Cl | Mn | Se | I |
| 125 | **Curry,** lentil, red/masoor dahl, mung bean dahl and tomato | N | 170 | 8 | 15 | 51 | (1.1) | 0.11 | 0.5 | N | N | (2) | N |
| 126 | lentil, red/masoor dahl and vegetable | 12 | 260 | 23 | 24 | 85 | 2.3 | 0.15 | 0.8 | 30 | N | (1) | N |
| 127 | lentil, whole/masoor, Gujerati | 440 | 340 | 27 | 35 | 100 | 3.1 | 0.30 | 1.1 | 710 | 0.4 | (26) | N |
| 128 | lentil, whole/masoor, Punjabi | 390 | 280 | 28 | 32 | 87 | 3.1 | 0.26 | 1.0 | 610 | 0.4 | 23 | N |
| 129 | mung bean dahl, Bengali | 200 | 210 | 8 | 22 | 63 | 1.0 | 0.09 | 0.4 | 320 | 0.2 | (2) | N |
| 130 | mung bean dahl, Punjabi | N | 220 | 7 | 21 | 56 | (1.0) | 0.15 | 0.6 | N | 0.2 | (4) | N |
| 131 | mung bean dahl and spinach | 450 | 680 | 140 | 72 | 120 | 3.1 | 0.15 | 1.1 | 610 | 0.7 | (4) | N |
| 132 | mung bean dahl and tomato | N | 230 | 9 | 20 | 53 | (1.0) | 0.13 | 0.5 | N | 0.2 | (3) | N |
| 133 | mung bean, whole, Gujerati | 370 | 390 | 29 | 42 | 94 | 1.7 | 0.15 | 0.7 | 580 | 0.2 | 4 | N |
| 134 | mung bean, whole, Punjabi | 230 | 250 | 19 | 28 | 66 | 1.2 | 0.09 | 0.5 | 360 | 0.2 | 3 | N |
| 135 | mung bean, whole and turnip leaves | 440 | 190 | 69 | 29 | 62 | 2.9 | 0.13 | 0.6 | 670 | N | N | N |
| 136 | okra | 300 | 350 | 120 | 53 | 55 | 1.0 | 0.12 | 0.5 | 500 | N | (1) | N |
| 137 | pea and potato | 380 | 220 | 17 | 16 | 44 | 1.0 | 0.04 | 0.3 | 600 | 0.1 | N | N |
| 138 | pigeon pea dahl, *with butter* | 270 | 300 | 16 | 22 | 68 | 1.0 | 0.20 | 0.6 | 410 | 0.2 | N | N |
| 139 | *with vegetable oil* | 230 | 300 | 15 | 22 | 67 | 0.9 | 0.19 | 0.5 | 350 | 0.2 | N | N |
| 140 | pigeon pea dahl and tomato | 180 | 170 | 5 | 11 | 34 | 0.4 | 0.10 | 0.3 | 270 | 0.1 | N | N |

| No. 15- | Food | Retinol µg | Carotene µg | Vitamin D µg | Vitamin E mg | Thiamin mg | Ribo-flavin mg | Niacin mg | Trypt/60 mg | Vitamin B6 mg | Vitamin B12 µg | Folate µg | Panto-thenate mg | Biotin µg | Vitamin C mg |
|---|---|---|---|---|---|---|---|---|---|---|---|---|---|---|---|
| 125 | **Curry**, lentil, red/masoor dahl, mung bean dahl and tomato | 0 | (54) | 0 | (1.05) | (0.07) | (0.03) | (0.3) | 0.7 | (0.07) | 0 | (11) | (0.22) | N | 1 |
| 126 | lentil, red/masoor dahl and vegetable | 0 | 685 | 0 | (1.18) | 0.12 | 0.04 | 0.6 | 0.8 | 0.19 | 0 | 9 | 0.31 | N | 11 |
| 127 | lentil, whole/masoor, Gujerati | 48 | 99 | Tr | 0.46 | 0.11 | 0.06 | 0.7 | 0.9 | 0.23 | Tr | 16 | N | N | 2 |
| 128 | lentil, whole/masoor, Punjabi | 0 | 115 | 0 | 2.01 | 0.09 | 0.05 | 0.6 | 0.8 | 0.19 | 0 | 13 | N | N | 1 |
| 129 | mung bean dahl, Bengali | 30 | 110 | Tr | 0.09 | (0.05) | (0.04) | (0.3) | 0.6 | (0.05) | Tr | (10) | (0.24) | N | Tr |
| 130 | mung bean dahl, Punjabi | 0 | N | 0 | 0.68 | (0.08) | (0.06) | (0.4) | 1.0 | (0.06) | 0 | (28) | (0.33) | N | 1 |
| 131 | mung bean dahl and spinach | 0 | 2935 | 0 | 1.77 | (0.11) | (0.10) | (1.2) | 1.4 | (0.17) | 0 | (71) | (0.47) | N | 10 |
| 132 | mung bean dahl and tomato | 0 | 170 | 0 | 0.89 | (0.08) | (0.05) | (0.5) | 0.9 | (0.07) | 0 | (25) | (0.31) | N | 1 |
| 133 | mung bean, whole, Gujerati | 0 | 74 | 0 | 1.64 | 0.09 | 0.05 | 0.6 | 1.0 | 0.11 | 0 | 18 | 0.40 | N | 2 |
| 134 | mung bean, whole, Punjabi | 24 | 110 | Tr | 0.18 | 0.06 | 0.04 | 0.4 | 0.7 | 0.07 | Tr | 13 | 0.28 | N | 1 |
| 135 | mung bean, whole and turnip leaves | 57 | (3275) | Tr | (3.39) | 0.06 | 0.11 | 0.5 | 0.8 | 0.09 | Tr | 34 | 0.21 | N | 11 |
| 136 | okra | 0 | 490 | 0 | N | 0.15 | 0.04 | 0.9 | 0.4 | 0.18 | 0 | 32 | 0.21 | N | 9 |
| 137 | pea and potato | 0 | 355 | 0 | 2.84 | 0.09 | 0.03 | 0.7 | 0.5 | 0.14 | 0 | 9 | (0.13) | 0.3 | 3 |
| 138 | pigeon pea dahl, *with butter* | 39 | 99 | Tr | 0.12 | 0.14 | 0.02 | 0.4 | 0.4 | N | Tr | 10 | N | N | Tr |
| 139 | *with vegetable oil* | 0 | 78 | 0 | 1.20 | 0.14 | 0.02 | 0.4 | 0.4 | N | 0 | 10 | N | N | Tr |
| 140 | pigeon pea dahl and tomato | 0 | 80 | 0 | 0.56 | 0.08 | 0.01 | 0.3 | 0.2 | N | 0 | 5 | N | N | 1 |

| No. 15- | Food | Description and main data sources |
|---|---|---|
| 141 | **Curry**, pigeon pea dahl with tomatoes and peanuts | Gujerati dish. Recipe from personal collection. Split pigeon peas |
| 142 | potato, Gujerati | Recipe from review of recipe collection |
| 143 | potato, Punjabi | Recipe from review of recipe collection |
| 144 | potato and pea | Bangladeshi dish. Recipe adapted from Carlson and de Wet, 1991 |
| 145 | red kidney bean, Gujerati | Recipe from review of recipe collection. Thin-medium consistency |
| 146 | red kidney bean, Punjabi | Recipe from dietary survey records. Medium consistency |
| 147 | red kidney and mung bean, whole | Recipe from review of recipe collection |
| 148 | spinach | Islami dish. Spinach and tomato. Recipe from the Aga Khan Health Board for the United Kingdom |
| 149 | spinach and potato | Frozen spinach and potato in masala sauce. Recipe from review of recipe collection |
| 150 | tinda and potato | Punjabi dish. Tinda gourd, potato, tomato and onion. Recipe from dietary survey records |
| 151 | vegetable, frozen mixed vegetables | Frozen mixed vegetables and potato. Recipe from review of recipe collection |
| 152 | vegetable, in sweet sauce | Vegetables in a basic UK-type curry sauce. Recipe from review of recipe collection |
| 153 | vegetable, Islami | Recipe adapted from a recipe from the Aga Khan Health Board for the United Kingdom. Assorted vegetables |
| 154 | vegetable, Pakistani | Recipe from a personal collection. Assorted vegetables, includes coconut |
| 155 | vegetable, retail, with rice | 4 samples, 2 brands; cooked in conventional and microwave ovens according to packet directions |
| 156 | vegetable, takeaway | 12 samples purchased from take-away outlets; curry only, no rice |
| 157 | vegetable, West Indian | Aubergine, okra, spinach and potato. Recipe from dietary survey records |
| 158 | vegetable, with yogurt | Basic UK-type curry. Mixed vegetables with yogurt. Recipe from Leeds Polytechnic |

| No. Food | Water | Total nitrogen | Protein | Fat | Carbohydrate | Energy value | |
|---|---|---|---|---|---|---|---|
| **15-** | g | g | g | g | g | kcal | kJ |
| 141 **Curry**, pigeon pea dahl with tomatoes and peanuts | 69.4 | 1.12 | 6.9 | 3.3 | 16.7 | 119 | 503 |
| 142 potato, Gujerati | 78.6 | 0.23 | 1.4 | 5.3 | 12.2 | 96 | 403 |
| 143 potato, Punjabi | 77.5 | 0.30 | 1.9 | 4.0 | 13.6 | 92 | 390 |
| 144 potato and pea | 76.3 | 0.47 | 2.9 | 3.8 | 13.0 | 92 | 387 |
| 145 red kidney bean, Gujerati | 74.9 | 0.64 | 4.0 | 6.1 | 9.8 | 106 | 445 |
| 146 red kidney bean, Punjabi | 74.8 | 0.75 | 4.7 | 5.6 | 10.1 | 106 | 448 |
| 147 red kidney and mung bean, whole | 73.1 | 0.62 | 3.9 | 12.1 | 7.6 | 152 | 634 |
| 148 spinach | 78.0 | 0.42 | 2.6 | 9.2 | 4.6 | 107 | 446 |
| 149 spinach and potato | 76.6 | 0.33 | 2.0 | 8.4 | 7.7 | 110 | 459 |
| 150 tinda and potato | 83.5 | 0.27 | 1.7 | 3.9 | 8.3 | 71 | 300 |
| 151 vegetable, frozen mixed vegetables | 80.3 | 0.40 | 2.5 | 6.1 | 6.9 | 88 | 368 |
| 152 vegetable, in sweet sauce | 87.3 | 0.23 | 1.4 | 2.1 | 6.7 | 49 | 208 |
| 153 vegetable, Islami | 80.2 | 0.42 | 2.6 | 4.5 | 8.4 | 79 | 330 |
| 154 vegetable, Pakistani | 82.1 | 0.36 | 2.2 | 2.6 | 8.7 | 60 | 251 |
| 155 vegetable, retail, with rice | 74.5 | 0.53 | 3.3 | 3.0 | 16.4 | 102 | 429 |
| 156 vegetable, takeaway | 79.3 | 0.39 | 2.5 | 7.4 | 7.6 | 105 | 438 |
| 157 vegetable, West Indian | 83.6 | 0.37 | 2.4 | 4.0 | 3.9 | 59 | 248 |
| 158 vegetable, with yogurt | 85.3 | 0.41 | 2.6 | 4.1 | 4.6 | 62 | 260 |

**Carbohydrate fractions, g per 100g food**

| No. 15- | Food | Starch | Total sugars | Glucose | Fructose | Galactose | Sucrose | Maltose | Lactose | Oligo-saccharides |
|---|---|---|---|---|---|---|---|---|---|---|
| | | | | | | Individual sugars | | | | |
| 141 | **Curry**, pigeon pea dahl with tomatoes and peanuts | 15.1 | 1.0 | 0.2 | 0.2 | 0 | 0.6 | 0 | 0 | 0.6 |
| 142 | potato, Gujerati | 9.7 | 2.4 | 0.3 | 0.2 | 0 | 2.0 | 0 | 0 | Tr |
| 143 | potato, Punjabi | 11.0 | 2.3 | 0.9 | 0.8 | 0 | 0.6 | 0 | Tr | 0.4 |
| 144 | potato and pea | 7.6 | 3.6 | 1.1 | 0.8 | 0 | 1.7 | 0 | 0 | 1.7 |
| 145 | red kidney bean, Gujerati | 6.4 | 2.6 | 0.3 | 0.3 | 0 | 2.0 | 0 | 0 | 0.8 |
| 146 | red kidney bean, Punjabi | 7.8 | 1.2 | 0.3 | 0.3 | 0 | 0.5 | 0 | 0 | 1.1 |
| 147 | red kidney and mung bean, whole | 6.6 | 0.3 | Tr | Tr | 0 | 0.3 | 0 | Tr | 0.6 |
| 148 | spinach | 0.3 | 3.9 | 1.6 | 1.6 | 0 | 0.6 | 0 | 0 | 0.4 |
| 149 | spinach and potato | 2.1 | 4.5 | 1.8 | 1.6 | 0 | 1.1 | 0 | 0 | 1.0 |
| 150 | tinda and potato | 4.6 | 3.3 | N | N | 0 | N | 0 | 0 | 0.5 |
| 151 | vegetable, frozen mixed vegetables | 3.8 | 2.9 | 0.6 | 0.5 | 0 | 1.8 | 0 | 0 | 0.3 |
| 152 | vegetable, in sweet sauce | 3.1 | 3.5 | 1.4 | 1.5 | 0 | 0.6 | 0 | 0 | 0.2 |
| 153 | vegetable, Islami | 4.0 | 3.8 | 1.7 | 1.5 | 0 | 0.5 | 0 | 0 | 0.5 |
| 154 | vegetable, Pakistani | 4.4 | 3.8 | 1.3 | 1.2 | 0 | 1.3 | 0 | 0 | 0.5 |
| 155 | vegetable, retail, with rice | 14.0 | 2.4 | 0.7 | 0.8 | 0 | 0.7 | Tr | 0.2 | Tr |
| 156 | vegetable, takeaway | 4.5 | 3.1 | (1.1) | (1.1) | 0 | 0.9 | 0 | 0 | Tr |
| 157 | vegetable, West Indian | 1.8 | 2.1 | 0.8 | 0.7 | 0 | 0.5 | 0 | 0 | Tr |
| 158 | vegetable, with yogurt | 0.3 | 3.9 | 1.2 | 1.0 | 0.3 | 0.8 | 0 | 0.5 | 0.4 |

Fibre fractions, phytic acid and fatty acids, g per 100g food
Cholesterol, mg per 100g food

| No. 15- | Food | Dietary fibre, g | | Fibre fractions, g | | | | Phytic acid g | Fatty acids, g | | | Cholesterol mg |
| | | Southgate method | Englyst method | Cellulose | Non-cellulosic polysaccharide | | Lignin | | Satd | Mono-unsatd | Poly-unsatd | |
| | | | | | Soluble | Insoluble | | | | | | | |
| 141 | **Curry**, pigeon pea dahl with tomatoes and peanuts | 3.4 | (3.0) | N | N | N | N | 0.30 | 0.5 | 1.2 | 1.3 | 0 |
| 142 | potato, Gujerati | 0.9 | 0.8 | 0.3 | 0.4 | 0.1 | Tr | Tr | 0.5 | 1.9 | 2.4 | 0 |
| 143 | potato, Punjabi | 1.3 | 1.2 | 0.4 | 0.6 | 0.1 | Tr | 0.01 | 2.5 | 1.0 | 0.2 | 10 |
| 144 | potato and pea | 3.1 | 2.4 | 1.1 | 1.1 | 0.3 | Tr | 0.02 | 0.4 | 1.2 | 1.7 | 0 |
| 145 | red kidney bean, Gujerati | (4.1) | 2.7 | 0.7 | 1.2 | 0.8 | N | 0.20 | 0.6 | 2.1 | 2.9 | 0 |
| 146 | red kidney bean, Punjabi | (5.0) | 3.8 | 1.0 | 1.8 | 1.0 | N | 0.15 | 0.6 | 1.9 | 2.7 | 0 |
| 147 | red kidney and mung bean, whole | (3.1) | 2.1 | 0.6 | 0.8 | 0.7 | N | 0.16 | 3.4 | 3.8 | 4.1 | 10 |
| 148 | spinach | (2.8) | 1.8 | 0.8 | 0.8 | 0.3 | Tr | 0.01 | 0.9 | 3.2 | 4.4 | 0 |
| 149 | spinach and potato | (2.5) | 1.8 | 0.7 | 0.8 | 0.3 | 0.2 | 0.03 | 0.9 | 2.8 | 4.0 | 0 |
| 150 | tinda and potato | 1.7 | N | N | N | N | N | N | 0.5 | 0.9 | 2.1 | 0 |
| 151 | vegetable, frozen mixed vegetables | N | N | N | N | N | Tr | 0.01 | 0.6 | 2.0 | 2.7 | 0 |
| 152 | vegetable, in sweet sauce | (1.6) | 1.3 | 0.5 | 0.6 | 0.3 | Tr | 0.02 | 0.6 | 0.5 | 0.8 | 0 |
| 153 | vegetable, Islami | 2.0 | 1.8 | 0.6 | 0.8 | 0.3 | Tr | 0.03 | 0.5 | 1.5 | 2.0 | 0 |
| 154 | vegetable, Pakistani | 2.7 | 2.2 | 0.9 | 0.9 | 0.4 | 0.1 | 0.01 | 0.8 | 0.6 | 0.9 | 0 |
| 155 | vegetable, retail, with rice | N | N | N | N | N | N | N | N | N | N | Tr |
| 156 | vegetable, takeaway | N | N | N | N | N | N | N | 1.5 | 3.7 | N | Tr |
| 157 | vegetable, West Indian | (3.9) | 3.1 | 1.0 | 1.5 | 0.5 | 0.3 | N | 0.5 | 1.2 | 1.2 | 0 |
| 158 | vegetable, with yogurt | 1.5 | 1.4 | 0.4 | 0.7 | 0.3 | Tr | 0.04 | 0.5 | 1.3 | 1.9 | Tr |

| No. 15- | Food | mg | | | | | | | | | | µg | |
|---|---|---|---|---|---|---|---|---|---|---|---|---|---|
| | | Na | K | Ca | Mg | P | Fe | Cu | Zn | Cl | Mn | Se | I |
| 141 | **Curry**, pigeon pea dahl with tomatoes and peanuts | 220 | 440 | 15 | 34 | 100 | 1.1 | 0.29 | 0.8 | 330 | 0.3 | N | N |
| 142 | potato, Gujerati | 330 | 220 | 14 | 16 | 26 | 0.8 | 0.06 | 0.3 | 540 | 0.1 | 1 | (3) |
| 143 | potato, Punjabi | 230 | 310 | 15 | 17 | 37 | 1.0 | 0.10 | 0.4 | 390 | 0.3 | 1 | (4) |
| 144 | potato and pea | 220 | 270 | 28 | 19 | 55 | 1.2 | (0.07) | 0.5 | 370 | 0.2 | (1) | 4 |
| 145 | red kidney bean, Gujerati | 190 | 280 | 23 | 29 | 77 | 1.4 | 0.13 | 0.6 | 290 | 0.2 | 3 | N |
| 146 | red kidney bean, Punjabi | N | 270 | 22 | 25 | 74 | (1.4) | 0.13 | 0.6 | N | 0.3 | 3 | N |
| 147 | red kidney and mung bean, whole | N | 220 | 16 | 25 | 65 | (1.1) | 0.10 | 0.5 | N | 0.2 | 3 | N |
| 148 | spinach | 290 | 490 | 110 | 41 | 48 | 1.9 | 0.10 | 0.5 | 410 | 0.4 | (1) | (5) |
| 149 | spinach and potato | 480 | 390 | 61 | 26 | 45 | 1.1 | 0.05 | 0.4 | 750 | 0.3 | (1) | (5) |
| 150 | tinda and potato | N | 210 | 21 | 14 | 35 | (0.8) | (0.09) | N | N | N | N | N |
| 151 | vegetable, frozen mixed vegetables | 300 | 180 | 30 | 17 | 45 | 1.0 | 0.05 | 0.4 | 470 | 0.2 | Tr | (2) |
| 152 | vegetable, in sweet sauce | 21 | 190 | 19 | 11 | 27 | 0.8 | 0.05 | 0.2 | 45 | 0.2 | Tr | (2) |
| 153 | vegetable, Islami | 29 | 400 | 29 | 21 | 50 | 1.1 | 0.10 | 0.4 | 82 | 0.2 | (1) | (3) |
| 154 | vegetable, Pakistani | 23 | 300 | 35 | 21 | 47 | 1.2 | (0.09) | 0.4 | 60 | 0.3 | (1) | 2 |
| 155 | vegetable, retail, with rice | 250 | 150 | 37 | 16 | 54 | 0.8 | 0.10 | 0.4 | 380 | 0.3 | N | N |
| 156 | vegetable, takeaway | 370 | N | N | N | N | 2.7 | N | N | 570 | N | N | N |
| 157 | vegetable, West Indian | 190 | 390 | 120 | 49 | 45 | 2.0 | 0.08 | 0.6 | 270 | N | N | N |
| 158 | vegetable, with yogurt | 91 | 310 | 47 | 18 | 60 | 0.9 | 0.04 | 0.4 | 160 | 0.2 | (1) | (8) |

| No. 15- | Food | Retinol µg | Carotene µg | Vitamin D µg | Vitamin E mg | Thiamin mg | Ribo-flavin mg | Niacin mg | Trypt 60 mg | Vitamin B6 mg | Vitamin B12 µg | Folate µg | Panto-thenate mg | Biotin µg | Vitamin C mg |
|---|---|---|---|---|---|---|---|---|---|---|---|---|---|---|---|
| 141 | **Curry**, pigeon pea dahl with tomatoes and peanuts | 0 | 80 | 0 | 0.75 | 0.23 | 0.03 | 1.0 | 0.7 | N | 0 | 16 | N | N | 1 |
| 142 | potato, Gujerati | 0 | 130 | 0 | 1.33 | 0.09 | 0.01 | 0.4 | 0.3 | 0.17 | 0 | 8 | 0.20 | 0.3 | 3 |
| 143 | potato, Punjabi | 37 | 160 | Tr | 0.50 | 0.13 | 0.02 | 0.6 | 0.4 | 0.23 | Tr | 11 | 0.27 | 0.6 | 5 |
| 144 | potato and pea | 0 | 320 | 0 | 1.00 | 0.17 | 0.03 | 0.9 | 0.5 | 0.20 | 0 | 18 | 0.19 | 0.5 | 5 |
| 145 | red kidney bean, Gujerati | 0 | 40 | 0 | 1.55 | 0.10 | 0.03 | 0.4 | 0.6 | 0.08 | 0 | 12 | 0.12 | N | 2 |
| 146 | red kidney bean, Punjabi | 0 | 76 | 0 | 1.54 | 0.11 | 0.03 | 0.4 | 0.7 | 0.09 | 0 | 23 | 0.15 | N | 2 |
| 147 | red kidney and mung bean, whole | 38 | 23 | Tr | (2.07) | 0.07 | 0.03 | 0.3 | 0.6 | 0.05 | Tr | 11 | 0.18 | N | Tr |
| 148 | spinach | 0 | 1880 | 0 | 3.86 | 0.08 | 0.05 | 1.0 | 0.5 | 0.16 | 0 | 41 | (0.24) | (1.1) | 11 |
| 149 | spinach and potato | 0 | 1360 | 0 | 3.05 | 0.12 | 0.03 | 1.0 | 0.5 | 0.21 | 0 | 29 | (0.23) | (1.0) | 9 |
| 150 | tinda and potato | 0 | 285 | 0 | N | 0.09 | 0.04 | 0.6 | 0.3 | N | 0 | N | N | N | 8 |
| 151 | vegetable, frozen mixed vegetables | 0 | 1495 | 0 | N | 0.09 | 0.05 | 0.6 | 0.4 | 0.12 | 0 | 17 | N | N | 5 |
| 152 | vegetable, in sweet sauce | 0 | 925 | 0 | 0.78 | 0.07 | 0.02 | 0.4 | 0.3 | 0.11 | 0 | 15 | 0.17 | (0.7) | 7 |
| 153 | vegetable, Islami | 0 | 165 | 0 | 1.70 | 0.12 | 0.03 | 0.7 | 0.5 | 0.22 | 0 | 17 | 0.29 | N | 10 |
| 154 | vegetable, Pakistani | 0 | (1775) | 0 | (0.67) | 0.12 | 0.03 | 0.6 | 0.4 | 0.16 | 0 | 16 | 0.18 | 0.6 | 6 |
| 155 | vegetable, retail, with rice | Tr | N | Tr | N | 0.06 | 0.08 | 0.6 | 0.7 | 0.13 | Tr | 8 | N | N | N |
| 156 | vegetable, takeaway | 0 | 640 | 0 | 0.92 | 0.03 | 0.05 | N | N | N | 0 | N | N | N | N |
| 157 | vegetable, West Indian | 0 | 1305 | 0 | N | 0.09 | 0.05 | 0.7 | 0.5 | 0.14 | 0 | 42 | (0.18) | N | 9 |
| 158 | vegetable, with yogurt | Tr | 1295 | Tr | (1.00) | 0.10 | 0.04 | 0.4 | 0.6 | 0.15 | Tr | 20 | 0.25 | N | 10 |

No. Food

**15-**

Description and main data sources

| No. | Food | Description and main data sources |
|---|---|---|
| 159 | **Dal Dhokari** | Gujerati dish. Pastry shapes in pigeon pea sauce. Recipe from a personal collection |
| 160 | **Dosa,** plain | Thin pancake made from rice and black gram dahl. Recipe from a personal collection |
| 161 | filling, vegetable | Filling for Dosa pancakes. Recipe from a personal collection |
| 162 | **Falafel,** *fried in vegetable oil* | Middle-Eastern dish. Deep fried chick pea balls. Recipe from review of recipe collection |
| 163 | **Flan,** broccoli | Recipe from dietary survey records |
| 164 | broccoli, wholemeal | Recipe from dietary survey records |
| 165 | cauliflower cheese | Recipe from dietary survey records and dissection of shop-bought samples |
| 166 | cauliflower cheese, wholemeal | Recipe from dietary survey records and dissection of shop-bought samples |
| 167 | cheese and mushroom | Recipe from dietary survey records |
| 168 | cheese and mushroom, wholemeal | |
| 169 | cheese, onion and potato | Recipe from dietary survey records |
| 170 | cheese, onion and potato, wholemeal | Recipe from dietary survey records |
| 171 | lentil and tomato | Recipe adapted from a recipe from Leeds Polytechnic |
| 172 | lentil and tomato, wholemeal | Recipe adapted from a recipe from Leeds Polytechnic |
| 173 | spinach | Recipe from review of recipe collection |
| 174 | spinach, wholemeal | Recipe from review of recipe collection |
| 175 | vegetable | Recipe from dietary survey records and dissection of shop-bought samples |
| 176 | vegetable, wholemeal | Recipe from dietary survey records and dissection of shop-bought samples |

**Vegetable Dishes**

Composition of food per 100g

| No. 15- | Food | Water g | Total nitrogen g | Protein g | Fat g | Carbohydrate g | Energy value | |
|---|---|---|---|---|---|---|---|---|
| | | | | | | | kcal | kJ |
| 159 | **Dal Dhokari** | 67.0 | 0.96 | 5.8 | 3.3 | 18.9 | 121 | 515 |
| 160 | **Dosa**, plain | 52.2 | 1.56 | 9.7 | 8.4 | 23.7 | 206 | 865 |
| 161 | filling, vegetable | 77.3 | 0.30 | 1.9 | 6.3 | 10.3 | 101 | 421 |
| 162 | **Falafel**, *fried in vegetable oil* | 60.0 | 1.03 | 6.4 | 11.2 | 15.6 | 179 | 750 |
| 163 | **Flan**, broccoli | 54.2 | 1.36 | 8.3 | 15.1 | 21.2 | 249 | 1039 |
| 164 | broccoli, wholemeal | 54.2 | 1.49 | 9.2 | 15.3 | 17.8 | 241 | 1006 |
| 165 | cauliflower cheese | 63.1 | 0.83 | 5.0 | 12.6 | 17.2 | 198 | 826 |
| 166 | cauliflower cheese, wholemeal | 63.1 | 0.93 | 5.7 | 12.8 | 14.6 | 192 | 801 |
| 167 | cheese and mushroom | 49.1 | 1.80 | 10.8 | 18.9 | 18.7 | 283 | 1182 |
| 168 | cheese and mushroom, wholemeal | 49.1 | 1.91 | 11.5 | 19.1 | 15.8 | 277 | 1154 |
| 169 | cheese, onion and potato | 40.6 | 2.09 | 13.1 | 23.9 | 20.3 | 343 | 1432 |
| 170 | cheese, onion and potato, wholemeal | 40.6 | 2.20 | 13.8 | 24.1 | 17.5 | 337 | 1404 |
| 171 | lentil and tomato | 60.9 | 1.22 | 7.5 | 6.3 | 22.8 | 171 | 724 |
| 172 | lentil and tomato, wholemeal | 60.9 | 1.30 | 8.0 | 6.4 | 20.9 | 167 | 705 |
| 173 | spinach | 61.0 | 1.60 | 10.0 | 13.0 | 13.1 | 205 | 859 |
| 174 | spinach, wholemeal | 61.0 | 1.68 | 10.5 | 13.1 | 11.1 | 201 | 840 |
| 175 | vegetable | 59.9 | 0.87 | 5.3 | 12.8 | 20.0 | 211 | 884 |
| 176 | vegetable, wholemeal | 59.9 | 0.99 | 6.0 | 13.0 | 17.0 | 204 | 855 |

# Vegetable Dishes

**Carbohydrate fractions, g per 100g food**

| No. 15- | Food | Starch | Total sugars | Individual sugars | | | | | | Oligo-saccharides |
|---|---|---|---|---|---|---|---|---|---|---|
| | | | | Glucose | Fructose | Galactose | Sucrose | Maltose | Lactose | |
| 159 | **Dal Dhokari** | 17.3 | 1.3 | (0.1) | (0.1) | 0 | (1.1) | 0 | 0 | 0.4 |
| 160 | **Dosa**, plain | 22.7 | 0.4 | Tr | Tr | 0 | 0.4 | 0 | 0 | 0.6 |
| 161 | filling, vegetable | 5.1 | 4.2 | 1.7 | 1.5 | 0 | 0.9 | 0 | 0 | 0.9 |
| 162 | **Falafel**, *fried in vegetable oil* | 11.4 | 2.6 | 0.8 | 0.6 | 0 | 1.2 | 0 | 0 | 1.6 |
| 163 | **Flan**, broccoli | 19.1 | 1.9 | 0.3 | 0.3 | 0 | 0.1 | Tr | 1.1 | Tr |
| 164 | broccoli, wholemeal | 15.5 | 2.1 | 0.3 | 0.3 | 0 | 0.3 | Tr | 1.1 | Tr |
| 165 | cauliflower cheese | 14.7 | 2.5 | 0.4 | 0.4 | 0 | 0.1 | Tr | 1.5 | Tr |
| 166 | cauliflower cheese, wholemeal | 11.9 | 2.6 | 0.4 | 0.4 | 0 | 0.2 | Tr | 1.5 | Tr |
| 167 | cheese and mushroom | 16.4 | 2.3 | 0.2 | 0.2 | 0 | 0.1 | Tr | 1.8 | Tr |
| 168 | cheese and mushroom, wholemeal | 13.3 | 2.4 | 0.2 | 0.2 | 0 | 0.3 | Tr | 1.8 | Tr |
| 169 | cheese, onion and potato | 19.3 | 0.9 | 0.3 | 0.2 | 0 | 0.2 | Tr | 0.2 | 0.1 |
| 170 | cheese, onion and potato, wholemeal | 16.3 | 1.0 | 0.3 | 0.2 | 0 | 0.4 | Tr | 0.2 | 0.1 |
| 171 | lentil and tomato | 19.7 | 2.3 | 0.8 | 0.7 | 0 | 0.7 | Tr | 0 | 0.9 |
| 172 | lentil and tomato, wholemeal | 17.6 | 2.4 | 0.8 | 0.7 | 0 | 0.8 | Tr | 0 | 0.9 |
| 173 | spinach | 11.3 | 1.5 | 0.3 | 0.2 | 0 | 0.3 | Tr | 0.6 | 0.2 |
| 174 | spinach, wholemeal | 9.2 | 1.6 | 0.3 | 0.2 | 0 | 0.4 | Tr | 0.6 | 0.2 |
| 175 | vegetable | 17.3 | 2.5 | 0.5 | 0.5 | 0 | 0.5 | Tr | 0.9 | 0.2 |
| 176 | vegetable, wholemeal | 14.1 | 2.7 | 0.6 | 0.5 | 0 | 0.7 | Tr | 0.9 | 0.2 |

| No. 15- | Food | Dietary fibre, g Southgate method | Dietary fibre, g Englyst method | Fibre fractions, g Cellulose | Non-cellulosic polysaccharide Soluble | Non-cellulosic polysaccharide Insoluble | Lignin | Phytic acid g | Fatty acids, g Satd | Mono-unsatd | Poly-unsatd | Cholesterol mg |
|---|---|---|---|---|---|---|---|---|---|---|---|---|
| 159 | **Dal Dhokari** | 3.1 | (3.0) | N | N | N | N | 0.24 | 0.5 | 1.1 | 1.4 | 0 |
| 160 | **Dosa**, plain | 4.3 | 3.7 | N | N | N | 0.5 | 0.34 | 0.9 | 2.9 | 4.0 | 0 |
| 161 | filling, vegetable | 1.7 | 1.4 | 0.5 | 0.7 | 0.2 | 0.2 | 0.03 | 0.7 | 2.1 | 2.8 | 0 |
| 162 | **Falafel**, *fried in vegetable oil* | 4.4 | 3.4 | 0.9 | 1.2 | 1.3 | 0.5 | 0.14 | 1.1 | 3.9 | 5.3 | 0 |
| 163 | **Flan**, broccoli | (1.3) | 1.2 | 0.2 | 0.5 | 0.4 | Tr | 0.03 | 5.9 | 5.2 | 2.7 | 86 |
| 164 | broccoli, wholemeal | (2.6) | 2.7 | 0.5 | 0.6 | 1.5 | 0.1 | 0.12 | 5.9 | 5.2 | 2.8 | 86 |
| 165 | cauliflower cheese | 1.2 | 1.1 | 0.2 | 0.5 | 0.4 | Tr | 0.04 | 5.1 | 4.2 | 2.4 | 30 |
| 166 | cauliflower cheese, wholemeal | 2.1 | 2.2 | 0.5 | 0.6 | 1.2 | 0.1 | 0.11 | 5.1 | 4.3 | 2.5 | 30 |
| 167 | cheese and mushroom | 1.2 | 0.9 | 0.1 | 0.4 | 0.5 | Tr | 0.04 | 7.7 | 5.9 | 4.1 | 106 |
| 168 | cheese and mushroom, wholemeal | 2.3 | 2.2 | 0.3 | 0.5 | 1.3 | 0.1 | 0.12 | 7.7 | 5.9 | 4.2 | 106 |
| 169 | cheese, onion and potato | 1.1 | 1.0 | 0.1 | 0.5 | 0.3 | Tr | 0.03 | 11.4 | 6.9 | 4.3 | 42 |
| 170 | cheese, onion and potato, wholemeal | 2.2 | 2.2 | 0.4 | 0.6 | 1.2 | 0.1 | 0.11 | 11.5 | 6.9 | 4.4 | 42 |
| 171 | lentil and tomato | 3.0 | 1.9 | 0.4 | 0.7 | 0.7 | 0.1 | 0.09 | 2.0 | 1.9 | 1.9 | 5 |
| 172 | lentil and tomato, wholemeal | 3.7 | 2.7 | 0.5 | 0.7 | 1.3 | 0.1 | 0.14 | 2.0 | 1.9 | 2.0 | 5 |
| 173 | spinach | (1.9) | (1.4) | (0.4) | (0.6) | (0.4) | Tr | 0.02 | 4.0 | 4.2 | 4.0 | 51 |
| 174 | spinach, wholemeal | (2.6) | (2.3) | (0.6) | (0.7) | (1.0) | (0.1) | 0.08 | 4.0 | 4.2 | 4.0 | 51 |
| 175 | vegetable | (1.7) | 1.5 | 0.3 | 0.8 | 0.4 | Tr | 0.03 | 4.3 | 3.9 | 4.0 | 10 |
| 176 | vegetable, wholemeal | (2.8) | 2.8 | 0.6 | 0.9 | 1.3 | 0.1 | 0.11 | 4.3 | 3.9 | 4.0 | 10 |

# Vegetable Dishes

## Inorganic constituents per 100g food

| No. 15- | Food | Na | K | Ca | Mg | P | Fe | Cu | Zn | Cl | Mn | Se | I |
|---|---|---|---|---|---|---|---|---|---|---|---|---|---|
| | | | | | | mg | | | | | | µg | |
| 159 | **Dal Dhokari** | 340 | 330 | 19 | 41 | 110 | 1.9 | 0.24 | 0.9 | 520 | 0.7 | N | N |
| 160 | **Dosa**, plain | 410 | 290 | 27 | 47 | 110 | 2.2 | 0.26 | 1.1 | (590) | 0.5 | N | N |
| 161 | filling, vegetable | 440 | 310 | 21 | 17 | 39 | 0.9 | 0.06 | 0.3 | 720 | 0.2 | 1 | 4 |
| 162 | **Falafel**, *fried in vegetable oil* | 290 | 380 | 80 | 45 | 130 | 2.8 | 0.29 | 1.0 | 410 | 0.7 | (1) | N |
| 163 | **Flan**, broccoli | 190 | 150 | 150 | 17 | 140 | 1.1 | 0.07 | 0.8 | (300) | 0.2 | 4 | 20 |
| 164 | broccoli, wholemeal | 190 | 200 | 130 | 42 | 190 | 1.6 | 0.15 | 1.4 | (290) | 0.8 | 16 | N |
| 165 | cauliflower cheese | 510 | 120 | 100 | 15 | 92 | 0.6 | 0.04 | 0.5 | 810 | 0.2 | 2 | 12 |
| 166 | cauliflower cheese, wholemeal | 510 | 160 | 84 | 34 | 130 | 1.0 | 0.10 | 1.0 | 800 | 0.7 | 11 | (10) |
| 167 | cheese and mushroom | 550 | 170 | 210 | 18 | 200 | 1.1 | 0.13 | 1.0 | N | 0.2 | 7 | 29 |
| 168 | cheese and mushroom, wholemeal | 550 | 220 | 190 | 40 | 250 | 1.5 | 0.20 | 1.5 | N | 0.7 | 17 | (27) |
| 169 | cheese, onion and potato | 470 | 130 | 330 | 18 | 240 | 0.7 | 0.07 | 1.2 | 750 | 0.2 | 6 | 22 |
| 170 | cheese, onion and potato, wholemeal | 470 | 170 | 310 | 39 | 280 | 1.1 | 0.13 | 1.7 | 740 | 0.7 | 16 | N |
| 171 | lentil and tomato | 82 | 270 | 83 | 27 | 120 | 2.0 | 0.17 | 0.9 | 150 | 0.2 | 2 | 6 |
| 172 | lentil and tomato, wholemeal | 82 | 300 | 69 | 41 | 150 | 2.3 | 0.22 | 1.2 | 150 | 0.5 | 9 | (4) |
| 173 | spinach | 420 | 210 | 170 | 22 | 150 | 1.3 | 0.10 | 0.9 | 660 | 0.2 | (4) | (11) |
| 174 | spinach, wholemeal | 420 | 240 | 160 | 37 | 180 | 1.6 | 0.15 | 1.3 | 650 | 0.6 | (11) | N |
| 175 | vegetable | 310 | 120 | 120 | 12 | 93 | 0.8 | 0.06 | 0.5 | 490 | 0.2 | 2 | 12 |
| 176 | vegetable, wholemeal | 310 | 160 | 98 | 33 | 140 | 1.2 | 0.12 | 1.0 | 480 | 0.7 | 13 | (10) |

| No. 15- | Food | Retinol µg | Carotene µg | Vitamin D µg | Vitamin E mg | Thiamin mg | Ribo-flavin mg | Niacin mg | Trypt 60 mg | Vitamin B6 mg | Vitamin B12 µg | Folate µg | Panto-thenate mg | Biotin µg | Vitamin C mg |
|---|---|---|---|---|---|---|---|---|---|---|---|---|---|---|---|
| 159 | **Dal Dhokari** | 0 | 26 | 0 | (0.82) | 0.17 | 0.03 | 1.2 | 0.8 | (0.07) | 0 | 13 | (0.15) | (1.8) | 1 |
| 160 | **Dosa**, plain | 0 | 135 | 0 | (1.88) | 0.11 | 0.11 | 0.8 | 1.9 | N | 0 | 22 | N | N | Tr |
| 161 | filling, vegetable | 0 | 395 | 0 | 2.09 | 0.12 | 0.01 | 0.8 | 0.4 | 0.20 | 0 | 11 | 0.23 | 1.0 | 6 |
| 162 | **Falafel**, *fried in vegetable oil* | 0 | 230 | 0 | 3.16 | 0.13 | 0.07 | 0.8 | 0.9 | 0.16 | 0 | 26 | 0.45 | N | 8 |
| 163 | **Flan**, broccoli | 155 | 200 | 1.1 | (1.39) | 0.08 | 0.14 | 0.6 | 2.0 | 0.08 | 0.6 | 14 | (0.35) | (4.1) | 4 |
| 164 | broccoli, wholemeal | 155 | 200 | 1.1 | (1.67) | 0.11 | 0.15 | 1.6 | 2.1 | 0.14 | 0.6 | 18 | (0.45) | (5.6) | 4 |
| 165 | cauliflower cheese | 125 | 125 | 0.8 | 0.98 | 0.07 | 0.08 | 0.5 | 1.1 | 0.08 | 0.2 | 11 | 0.25 | 1.5 | 4 |
| 166 | cauliflower cheese, wholemeal | 125 | 125 | 0.8 | 1.19 | 0.09 | 0.09 | 1.2 | 1.2 | 0.13 | 0.2 | 15 | 0.32 | 2.6 | 4 |
| 167 | cheese and mushroom | 205 | 130 | 1.3 | 0.46 | 0.09 | 0.26 | 0.8 | 2.6 | 0.08 | 0.9 | 12 | 0.66 | 7.2 | Tr |
| 168 | cheese and mushroom, wholemeal | 205 | 130 | 1.3 | 0.69 | 0.11 | 0.27 | 1.6 | 2.7 | 0.14 | 0.9 | 16 | 0.74 | 8.5 | Tr |
| 169 | cheese, onion and potato | 230 | 180 | 1.0 | 0.31 | 0.09 | 0.15 | 0.5 | 3.0 | 0.12 | 0.5 | 12 | 0.22 | 1.6 | 1 |
| 170 | cheese, onion and potato, wholemeal | 230 | 180 | 1.0 | 0.54 | 0.12 | 0.16 | 1.3 | 3.1 | 0.17 | 0.5 | 16 | 0.30 | 2.8 | 1 |
| 171 | lentil and tomato | 44 | 140 | 0.3 | (0.61) | 0.14 | 0.05 | 0.8 | 1.3 | 0.16 | 0.1 | 9 | 0.33 | (0.8) | 3 |
| 172 | lentil and tomato, wholemeal | 44 | 140 | 0.3 | (0.77) | 0.16 | 0.06 | 1.4 | 1.4 | 0.20 | 0.1 | 12 | 0.39 | (1.7) | 3 |
| 173 | spinach | 110 | (1540) | 0.8 | 1.49 | 0.08 | 0.14 | 0.6 | 2.4 | 0.09 | 0.6 | 26 | 0.37 | 2.8 | 1 |
| 174 | spinach, wholemeal | 110 | (1540) | 0.8 | 1.65 | 0.10 | 0.14 | 1.0 | 2.5 | 0.14 | 0.6 | 29 | 0.43 | 3.5 | 1 |
| 175 | vegetable | 120 | 1345 | 1.0 | (0.48) | 0.09 | 0.06 | 0.5 | 1.1 | 0.08 | 0.2 | 10 | (0.14) | (0.8) | 4 |
| 176 | vegetable, wholemeal | 120 | 1345 | 1.0 | (0.71) | 0.11 | 0.07 | 1.3 | 1.3 | 0.13 | 0.2 | 14 | (0.22) | (2.1) | 4 |

| No. | Food | Description and main data sources |
|-----|------|-----------------------------------|
| **15-** | | |
| 177 | **Fu-fu,** sweet potato | West Indian/African dish. Sweet potato and rice flour. Recipe adapted from Gillard et al, 1991 |
| 178 | yam | West Indian/African dish. Yam and rice flour. Recipe adapted from Gillard et al, 1991 |
| 179 | **Garlic mushrooms** | Recipe from review of recipe collection |
| 180 | **Guacamole** | Mashed avocado with tomato. Recipe from review of recipe collection |
| 181 | **Khadhi** | Gujerati dish. Yogurt with chick pea flour and spices. Recipe from a personal collection |
| 182 | **Khatiyu** | Gujerati dish. Whole pigeon peas with jaggery and spices. Recipe from a personal collection |
| 183 | **Khichadi,** with butter ghee | Mung bean and rice dish. Recipe from a personal collection |
| 184 | with vegetable oil | Mung bean and rice dish. Recipe from a personal collection |
| 185 | **Lasagne,** spinach | Recipe from dietary survey records |
| 186 | spinach, wholemeal | Recipe from dietary survey records |
| 187 | vegetable | Recipe from dietary survey records |
| 188 | vegetable, wholemeal | Recipe from dietary survey records |
| 189 | vegetable, retail | 8 assorted samples; cooked in conventional and microwave ovens according to packet directions |
| 190 | **Laverbread** | 6 samples; cooked pureed seaweed coated in oatmeal |
| 191 | **Leeks in cheese sauce,** made with whole milk | Recipe from review of recipe collection |
| 192 | made with semi-skimmed milk | |
| 193 | made with skimmed milk | Recipe from review of recipe collection |
| 194 | **Lentil cutlets,** fried in vegetable oil | Gujerati dish. Red lentil and potato cutlets. Recipe from a personal collection |

# Vegetable Dishes

## Composition of food per 100g

| No. 15- | Food | Water g | Total nitrogen g | Protein g | Fat g | Carbohydrate g | Energy value kcal | kJ |
|---|---|---|---|---|---|---|---|---|
| 177 | **Fu-fu**, sweet potato | 69.0 | 0.26 | 1.6 | 0.4 | 26.5 | 108 | 464 |
| 178 | yam | 59.5 | 0.35 | 2.1 | 0.4 | 38.0 | 154 | 658 |
| 179 | **Garlic mushrooms** | 77.6 | 0.70 | 2.1 | 14.4 | 0.6 | 139 | 574 |
| 180 | **Guacamole** | 79.3 | 0.23 | 1.4 | 12.7 | 2.2 a | 128 | 530 |
| 181 | **Khadhi** | 85.3 | 0.49 | 3.1 | 1.9 | 6.7 | 54 | 228 |
| 182 | **Khatiyu** | 54.0 | 1.35 | 8.4 | 7.0 | 26.0 | 193 | 816 |
| 183 | **Khichadi**, *with butter ghee* | 61.0 | 0.79 | 4.8 | 5.2 | 27.3 | 168 | 711 |
| 184 | *with vegetable oil* | 61.0 | 0.79 | 4.8 | 5.2 | 27.3 | 168 | 711 |
| 185 | **Lasagne**, spinach | 80.0 | 0.57 | 3.5 | 3.0 | 12.5 | 87 | 367 |
| 186 | spinach, wholemeal | 77.0 | 0.70 | 4.2 | 3.1 | 13.0 | 93 | 395 |
| 187 | vegetable | 77.7 | 0.68 | 4.1 | 4.4 | 12.4 | 102 | 431 |
| 188 | vegetable, wholemeal | 75.5 | 0.80 | 4.8 | 4.6 | 12.4 | 106 | 447 |
| 189 | vegetable, retail | 72.6 | 0.77 | 4.8 | 5.3 | 13.4 | 117 | 492 |
| 190 | **Laverbread** | 87.7 | 0.51 | 3.2 | 3.7 | 1.6 | 52 | 217 |
| 191 | **Leeks in cheese sauce**, *made with whole milk* | 81.0 | 0.70 | 4.4 | 7.0 | 5.1 | 99 | 412 |
| 192 | *made with semi-skimmed milk* | 81.7 | 0.71 | 4.4 | 6.1 | 5.2 | 92 | 384 |
| 193 | *made with skimmed milk* | 82.1 | 0.71 | 4.4 | 5.6 | 5.2 | 87 | 365 |
| 194 | **Lentil cutlets**, *fried in vegetable oil* | 54.3 | 1.33 | 8.3 | 8.1 | 22.9 | 189 | 794 |

a Including mannoheptulose present in avocados

# Vegetable Dishes

Carbohydrate fractions, g per 100g food

| No. 15- | Food | Starch | Total sugars | Individual sugars | | | | | | Oligo-saccharides |
|---|---|---|---|---|---|---|---|---|---|---|
| | | | | Glucose | Fructose | Galactose | Sucrose | Maltose | Lactose | |
| 177 | **Fu-fu**, sweet potato | 15.9 | 10.6 | N | N | 0 | N | N | 0 | 0 |
| 178 | yam | 37.4 | 0.6 | 0.2 | 0.1 | 0 | 0.4 | 0 | 0 | 0 |
| 179 | **Garlic mushrooms** | 0.3 | 0.2 | 0.1 | 0.1 | 0 | 0.1 | 0 | Tr | 0 |
| 180 | **Guacamole** | Tr | 1.3 [a] | 0.6 | 0.6 | 0 | 0.1 | 0 | 0 | 0 |
| 181 | **Khadhi** | 1.9 | 4.7 | Tr | Tr | 1.2 | 1.4 | 0 | 2.0 | 0.1 |
| 182 | **Khatiyu** | 19.9 | 5.6 | N | N | 0 | N | 0 | 0 | 0.5 |
| 183 | **Khichadi**, *with butter ghee* | 26.6 | 0.2 | Tr | Tr | 0 | 0.1 | 0 | Tr | 0.5 |
| 184 | *with vegetable oil* | 26.6 | 0.2 | Tr | Tr | 0 | 0.1 | 0 | 0 | 0.5 |
| 185 | **Lasagne**, spinach | 10.7 | 1.8 | 0.4 | 0.4 | 0 | 0.1 | 0.1 | 0.7 | 0 |
| 186 | spinach, wholemeal | 10.9 | 2.2 | 0.5 | 0.4 | 0 | 0.3 | 0.2 | 0.7 | 0 |
| 187 | vegetable | 9.7 | 2.7 | 0.7 | 0.7 | 0 | 0.2 | 0.1 | 1.0 | Tr |
| 188 | vegetable, wholemeal | 9.4 | 3.0 | 0.8 | 0.7 | 0 | 0.3 | 0.1 | 1.0 | Tr |
| 189 | vegetable, retail | 9.0 | 4.4 | 0.7 | 0.7 | 0 | 0.2 | 1.6 | 1.2 | Tr |
| 190 | **Laverbread** | 1.6 | Tr | Tr | Tr | 0 | Tr | 0 | 0 | 0 |
| 191 | **Leeks in cheese sauce**, *made with whole milk* | 2.0 | 2.9 | 0.4 | 0.5 | 0 | 0.3 | Tr | 1.7 | 0.2 |
| 192 | *made with semi-skimmed milk* | 2.0 | 3.0 | 0.4 | 0.5 | 0 | 0.3 | Tr | 1.8 | 0.2 |
| 193 | *made with skimmed milk* | 2.0 | 3.0 | 0.4 | 0.5 | 0 | 0.3 | Tr | 1.8 | 0.2 |
| 194 | **Lentil cutlets**, *fried in vegetable oil* | 20.8 | 1.2 | 0.1 | 0.2 | 0 | 0.8 | 0.1 | 0 | 1.0 |

[a] Not including mannoheptulose present in avocados

| No. 15- | Food | Dietary fibre, g Southgate method | Englyst method | Cellulose | Non-cellulosic polysaccharide Soluble | Insoluble | Lignin | Phytic acid g | Satd | Mono- unsatd | Poly- unsatd | Cholesterol mg |
|---|---|---|---|---|---|---|---|---|---|---|---|---|
| 177 | **Fu-fu**, sweet potato | 2.1 | 2.1 | 0.9 | 1.0 | 0.3 | 0.2 | 0.05 | 0.1 | 0.1 | 0.1 | 0 |
| 178 | yam | 3.4 | 1.3 | 0.6 | 0.5 | 0.1 | 0.2 | 0.04 | Tr | Tr | Tr | 0 |
| 179 | **Garlic mushrooms** | 2.5 | 1.2 | 0.2 | 0.2 | 0.7 | Tr | 0.09 | 9.2 | 3.4 | 0.8 | 38 |
| 180 | **Guacamole** | (2.6) | 2.5 | 0.8 | 1.1 | 0.5 | 0.2 | 0.01 | 2.7 | 7.9 | 1.5 | 0 |
| 181 | **Khadhi** | 0.6 | 0.5 | 0.1 | 0.1 | 0.2 | 0.1 | 0.04 | 1.1 | 0.5 | 0.2 | 5 |
| 182 | **Khatiyu** | 6.0 | (4.9) | N | N | N | 1.0 | 0.29 | 0.8 | 2.3 | 3.4 | 0 |
| 183 | **Khichadi**, *with butter ghee* | 2.4 | 1.4 | 0.5 | 0.4 | 0.5 | N | 0.20 | 3.0 | 1.3 | 0.5 | 11 |
| 184 | *with vegetable oil* | 2.4 | 1.4 | 0.5 | 0.4 | 0.5 | N | 0.20 | 0.7 | 1.7 | 2.4 | 0 |
| 185 | **Lasagne**, spinach | 1.6 | (1.1) | (0.3) | (0.5) | (0.3) | Tr | N | 1.3 | 0.8 | 0.5 | 6 |
| 186 | spinach, wholemeal | 2.8 | (2.3) | (0.6) | (0.6) | (1.0) | 0.1 | N | 1.3 | 0.8 | 0.6 | 6 |
| 187 | vegetable | 1.3 | (1.0) | (0.3) | (0.4) | (0.2) | Tr | 0.01 | 2.2 | 1.2 | 0.7 | 9 |
| 188 | vegetable, wholemeal | 2.4 | (2.1) | (0.5) | (0.6) | (0.9) | 0.1 | 0.02 | 2.2 | 1.2 | 0.8 | 9 |
| 189 | vegetable, retail | N | N | N | N | N | N | N | N | N | N | N |
| 190 | **Laverbread** | 2.8 | N | N | N | N | N | N | N | N | N | 0 |
| 191 | **Leeks in cheese sauce**, *made with whole milk* | 1.5 | 1.1 | 0.4 | 0.6 | 0.2 | 0.1 | 0.02 | 3.5 | 2.0 | 0.9 | 17 |
| 192 | *made with semi-skimmed milk* | 1.5 | 1.1 | 0.4 | 0.6 | 0.2 | 0.1 | 0.02 | 3.0 | 1.8 | 0.9 | 15 |
| 193 | *made with skimmed milk* | 1.5 | 1.1 | 0.4 | 0.6 | 0.2 | 0.1 | 0.02 | 2.7 | 1.6 | 0.9 | 13 |
| 194 | **Lentil cutlets**, *fried in vegetable oil* | 3.6 | 1.7 | 0.4 | 0.5 | 0.9 | 0.2 | 0.08 | 0.8 | 2.8 | 3.8 | 0 |

# Vegetable Dishes

**Inorganic constituents per 100g food**

| No. 15- | Food | Na | K | Ca | Mg | P | Fe | Cu | Zn | Cl | Mn | Se | I |
|---|---|---|---|---|---|---|---|---|---|---|---|---|---|
| | | | | | | mg | | | | | | µg | |
| 177 | **Fu-fu**, sweet potato | 29 | 280 | 21 | 42 | 54 | 0.7 | 0.14 | 0.4 | 49 | 0.4 | (2) | (3) |
| 178 | yam | 15 | 250 | 11 | 12 | 28 | 0.4 | 0.04 | 0.5 | 38 | 0.1 | N | N |
| 179 | **Garlic mushrooms** | 130 | 350 | 9 | 10 | 90 | 0.7 | 0.77 | 0.5 | 270 | 0.1 | 10 | 10 |
| 180 | **Guacamole** | 140 | 370 | 9 | 19 | 32 | 0.4 | 0.13 | 0.3 | 220 | 0.2 | Tr | 2 |
| 181 | **Khadhi** | 530 | 160 | 92 | 18 | 85 | 0.7 | 0.03 | 0.4 | 830 | 0.1 | 1 | 27 |
| 182 | **Khatiyu** | 310 | 510 | 61 | 45 | 120 | 1.9 | 0.45 | 1.0 | 460 | 0.5 | N | N |
| 183 | **Khichadi**, with butter ghee | 490 | 200 | 24 | 30 | 83 | 1.0 | 0.16 | 0.8 | 760 | 0.4 | 5 | N |
| 184 | with vegetable oil | 490 | 200 | 24 | 30 | 83 | 1.0 | 0.16 | 0.8 | 750 | 0.4 | 5 | N |
| 185 | **Lasagne**, spinach | 110 | 150 | 82 | 18 | 63 | 0.7 | 0.06 | 0.5 | 160 | 0.3 | (3) | (5) |
| 186 | spinach, wholemeal | 130 | 200 | 84 | 32 | 91 | 1.1 | 0.10 | 0.8 | 190 | 0.6 | N | (5) |
| 187 | vegetable | 98 | 180 | 85 | 14 | 82 | 0.5 | 0.09 | 0.5 | 160 | 0.2 | (3) | 7 |
| 188 | vegetable, wholemeal | 110 | 230 | 84 | 27 | 110 | 0.9 | 0.13 | 0.8 | 180 | 0.5 | N | N |
| 189 | vegetable, retail | 390 | 190 | 73 | 18 | 87 | 0.8 | 0.02 | 0.4 | 620 | 0.2 | N | N |
| 190 | **Laverbread** | 560 | 220 | 20 | 31 | 51 | 3.5 | 0.12 | 0.8 | 820 | N | N | N |
| 191 | **Leeks in cheese sauce**, made with whole milk | 200 | 150 | 120 | 8 | 98 | 0.5 | 0.02 | 0.5 | 330 | 0.1 | (2) | N |
| 192 | made with semi-skimmed milk | 200 | 150 | 120 | 8 | 99 | 0.5 | 0.02 | 0.5 | 330 | 0.1 | (2) | N |
| 193 | made with skimmed milk | 200 | 150 | 120 | 8 | 99 | 0.5 | 0.02 | 0.5 | 330 | 0.1 | (2) | N |
| 194 | **Lentil cutlets**, fried in vegetable oil | 770 | 330 | 33 | 41 | 120 | 3.0 | 0.22 | 1.1 | 1190 | N | (2) | N |

# Vegetable Dishes

**Vitamins per 100g food**

| No. 15- | Food | Retinol µg | Carotene µg | Vitamin D µg | Vitamin E mg | Thiamin mg | Riboflavin mg | Niacin mg | Trypt/60 mg | Vitamin B6 mg | Vitamin B12 µg | Folate µg | Pantothenate mg | Biotin µg | Vitamin C mg |
|---|---|---|---|---|---|---|---|---|---|---|---|---|---|---|---|
| 177 | **Fu-fu**, sweet potato | 0 | 3600 | 0 | 4.00 | 0.07 | 0.01 | 0.6 | 0.4 | 0.07 | 0 | 9 | 0.54 | N | 16 |
| 178 | yam | 0 | Tr | 0 | N | 0.13 | 0.01 | 0.3 | 0.5 | 0.14 | 0 | 7 | 0.34 | N | 4 |
| 179 | **Garlic mushrooms** | 135 | 72 | 0.1 | 0.47 | 0.08 | 0.27 | 2.7 | 0.3 | 0.15 | Tr | 23 | 1.70 | 10.1 | Tr |
| 180 | **Guacamole** | 0 | 190 | 0 | 2.42 | 0.09 | 0.12 | 1.0 | 0.2 | 0.28 | 0 | 12 | 0.79 | 2.8 | 11 |
| 181 | **Khadhi** | 11 | 16 | Tr | 0.14 | 0.04 | 0.11 | 0.2 | 0.6 | 0.06 | 0.1 | 12 | 0.23 | 1.2 | 1 |
| 182 | **Khatiyu** | 0 | 53 | 0 | 1.67 | 0.13 | 0.04 | 0.6 | 0.8 | N | 0 | 15 | N | N | Tr |
| 183 | **Khichadi**, with butter ghee | 28 | 23 | 0.1 | 0.16 | 0.12 | 0.03 | 1.1 | 0.9 | 0.10 | Tr | 11 | (0.31) | N | Tr |
| 184 | with vegetable oil | 0 | 3 | 0 | 1.03 | 0.12 | 0.03 | 1.1 | 0.9 | 0.10 | 0 | 11 | (0.31) | N | Tr |
| 185 | **Lasagne**, spinach | 25 | 950 | 0.1 | (0.79) | 0.05 | 0.06 | 0.6 | 0.8 | 0.06 | 0.1 | (26) | (0.15) | 0.7 | 5 |
| 186 | spinach, wholemeal | 25 | 950 | 0.1 | (0.79) | 0.13 | 0.06 | 1.0 | 0.9 | 0.09 | 0.1 | (27) | (0.24) | 0.7 | 5 |
| 187 | vegetable | 41 | 620 | 0.1 | (0.63) | 0.05 | 0.07 | 0.5 | 0.9 | 0.07 | 0.1 | (7) | (0.21) | (1.2) | 3 |
| 188 | vegetable, wholemeal | 41 | 620 | 0.1 | (0.63) | 0.10 | 0.07 | 0.8 | 1.0 | 0.10 | 0.1 | (8) | (0.28) | (1.4) | 3 |
| 189 | vegetable, retail | N | N | N | N | 0.25 | 0.29 | 2.0 | 1.1 | 0.44 | N | 6 | N | N | N |
| 190 | **Laverbread** | 0 | N | 0 | 1.10 | 0.03 | 0.10 | 0.6 | 0.5 | N | Tr | 47 | N | N | 5 |
| 191 | **Leeks in cheese sauce**, made with whole milk | 66 | 380 | 0.2 | 0.73 | 0.03 | 0.11 | 0.3 | 1.0 | 0.06 | 0.2 | 26 | 0.19 | 1.5 | 4 |
| 192 | made with semi-skimmed milk | 56 | 375 | 0.2 | 0.71 | 0.03 | 0.11 | 0.3 | 1.0 | 0.06 | 0.2 | 26 | 0.18 | 1.6 | 4 |
| 193 | made with skimmed milk | 49 | 370 | 0.2 | 0.70 | 0.03 | 0.11 | 0.3 | 1.0 | 0.06 | 0.2 | 26 | 0.18 | 1.5 | 4 |
| 194 | **Lentil cutlets**, fried in vegetable oil | 0 | 37 | 0 | 1.79 | 0.17 | 0.08 | 0.9 | 1.1 | 0.19 | 0 | 8 | 0.50 | N | 6 |

| No. | Food | Description and main data sources |
|-----|------|-----------------------------------|
| 15- | | |
| 195 | **Lentil pie** | Recipe from review of recipe collection |
| 196 | **Lentil and cheese pie** | Recipe from dietary survey records |
| 197 | **Lentil and potato pie** | Recipe from dietary survey records |
| 198 | **Lentil roast** | Recipe from review of recipe collection |
| 199 | *with egg* | Recipe from review of recipe collection |
| 200 | **Lentil and nut roast** | Recipe from review of recipe collection |
| 201 | *with egg* | Recipe from review of recipe collection |
| 202 | **Lentil and rice roast** | Recipe from dietary survey records |
| 203 | *with egg* | Recipe from review of recipe collection |
| 204 | **Mchicha** | West Indian dish. Steamed spinach with onion and tomato. Recipe from Gillard et al, 1991 |
| 205 | **Moussaka**, vegetable | Recipe from Leeds Polytechnic |
| 206 | vegetable, retail | 7 samples, 3 brands; cooked in conventional and microwave ovens according to packet directions |
| 207 | **Mushroom Dopiaza**, retail | Mushroom, onion, tomato and spices. Recipe from manufacturer |
| 208 | **Nut croquettes**, *fried in sunflower oil* | Recipe from review of recipe collection |
| 209 | *fried in vegetable oil* | Recipe from review of recipe collection |
| 210 | **Nut cutlets**, retail, *fried in sunflower oil* | Recipe from manufacturer |
| 211 | *fried in vegetable oil* | Recipe from manufacturer |
| 212 | *grilled* | Recipe from manufacturer |

# Vegetable Dishes

## Composition of food per 100g

| No. 15- | Food | Water g | Total nitrogen g | Protein g | Fat g | Carbohydrate g | Energy value kcal | Energy value kJ |
|---|---|---|---|---|---|---|---|---|
| 195 | **Lentil pie** | 59.4 | 1.18 | 7.2 | 4.4 | 23.5 | 156 | 659 |
| 196 | **Lentil and cheese pie** | 60.7 | 1.76 | 11.1 | 9.2 | 14.2 | 180 | 756 |
| 197 | **Lentil and potato pie** | 70.1 | 0.85 | 5.3 | 2.2 | 18.7 | 110 | 468 |
| 198 | **Lentil roast** | 61.8 | 1.25 | 7.7 | 3.0 | 21.9 | 139 | 588 |
| 199 | *with egg* | 62.5 | 1.33 | 8.2 | 3.6 | 20.3 | 141 | 596 |
| 200 | **Lentil and nut roast** | 51.6 | 1.83 | 10.6 | 12.1 | 18.8 | 222 | 929 |
| 201 | *with egg* | 52.9 | 1.85 | 10.8 | 12.1 | 17.6 | 218 | 914 |
| 202 | **Lentil and rice roast** | 65.8 | 0.76 | 4.7 | 1.9 | 26.6 | 135 | 572 |
| 203 | *with egg* | 66.1 | 0.82 | 5.0 | 2.3 | 25.5 | 136 | 576 |
| 204 | **Mchicha** | 79.4 | 0.59 | 3.7 | 4.8 | 4.0 | 73 | 303 |
| 205 | **Moussaka**, vegetable | 74.2 | 0.76 | 4.6 | 9.5 | 9.1 | 137 | 572 |
| 206 | vegetable, retail | 75.4 | 0.94 | 5.9 | 4.9 | 8.0 | 98 | 410 |
| 207 | **Mushroom Dopiaza**, retail | 86.0 | 0.33 | 1.3 | 5.7 | 3.7 | 69 | 286 |
| 208 | **Nut croquettes**, *fried in sunflower oil* | 36.2 | 1.72 | 9.3 | 26.2 | 21.6 | 351 | 1461 |
| 209 | *fried in vegetable oil* | 36.2 | 1.72 | 9.3 | 26.2 | 21.6 | 351 | 1461 |
| 210 | **Nut cutlets**, retail, *fried in sunflower oil* | 51.6 | 0.90 | 4.8 | 22.3 | 18.7 | 289 | 1205 |
| 211 | *fried in vegetable oil* | 51.6 | 0.90 | 4.8 | 22.3 | 18.7 | 289 | 1205 |
| 212 | *grilled* | 59.3 | 0.96 | 5.1 | 13.0 | 19.9 | 212 | 886 |

| No. 15- | Food | Starch | Total sugars | Individual sugars | | | | | | Oligo-saccharides |
|---|---|---|---|---|---|---|---|---|---|---|
| | | | | Glucose | Fructose | Galactose | Sucrose | Maltose | Lactose | |
| 195 | Lentil pie | 19.1 | 3.7 | N | N | 0 | N | N | 0 | 0.8 |
| 196 | Lentil and cheese pie | 12.0 | 1.4 | 0.4 | 0.3 | 0 | 0.7 | 0 | Tr | 0.8 |
| 197 | Lentil and potato pie | 16.6 | 1.4 | 0.4 | 0.3 | 0 | 0.7 | 0 | Tr | 0.6 |
| 198 | Lentil roast | 18.6 | 2.5 | N | N | 0 | N | N | 0 | 0.7 |
| 199 | with egg | 17.2 | 2.4 | N | N | 0 | N | N | 0 | 0.7 |
| 200 | Lentil and nut roast | 15.4 | 2.7 | N | N | 0 | N | N | 0 | 0.6 |
| 201 | with egg | 14.5 | 2.6 | N | N | 0 | N | N | 0 | 0.6 |
| 202 | Lentil and rice roast | 25.0 | 1.2 | (0.3) | (0.2) | 0 | (0.5) | Tr | 0 | 0.4 |
| 203 | with egg | 24.0 | 1.1 | (0.3) | (0.2) | 0 | (0.5) | Tr | 0 | 0.4 |
| 204 | Mchicha | 0.7 | 3.0 | 1.1 | 1.0 | 0 | 0.9 | 0 | 0 | 0.4 |
| 205 | Moussaka, vegetable | 6.2 | 2.6 | 0.7 | 0.6 | 0 | 0.4 | Tr | 1.0 | 0.3 |
| 206 | vegetable, retail | 3.5 | 4.5 | 1.0 | 1.0 | 0 | 0.5 | 0.9 | 1.1 | Tr |
| 207 | Mushroom Dopiaza, retail | 0.1 | 2.6 | 1.0 | 0.9 | 0 | 0.7 | 0 | 0 | 0.8 |
| 208 | Nut croquettes, fried in sunflower oil | 18.6 | 2.5 | N | N | 0 | N | N | 0 | 0.4 |
| 209 | fried in vegetable oil | 18.6 | 2.5 | N | N | 0 | N | N | 0 | 0.4 |
| 210 | Nut cutlets, retail, fried in sunflower oil | 17.3 | 1.3 | 0.3 | 0.2 | 0 | 0.7 | Tr | 0 | 0.1 |
| 211 | fried in vegetable oil | 17.3 | 1.3 | 0.3 | 0.2 | 0 | 0.7 | Tr | 0 | 0.1 |
| 212 | grilled | 18.4 | 1.4 | 0.3 | 0.2 | 0 | 0.8 | Tr | 0 | 0.1 |

**Vegetable Dishes**

## Fibre fractions, phytic acid and fatty acids, g per 100g food
## Cholesterol, mg per 100g food

| No. 15- | Food | Dietary fibre, g | | Fibre fractions, g | | | | Phytic acid g | Fatty acids, g | | | Cholesterol mg |
| | | Southgate method | Englyst method | Cellulose | Non-cellulosic polysaccharide | | Lignin | | Satd | Mono-unsatd | Poly-unsatd | |
| | | | | | Soluble | Insoluble | | | | | | |
| 195 | **Lentil pie** | (4.8) | (3.8) | 1.0 | 1.3 | 1.6 | N | 0.22 | 0.5 | 1.4 | 2.0 | 0 |
| 196 | **Lentil and cheese pie** | (2.9) | 2.4 | 0.7 | 0.7 | 1.0 | N | 0.10 | 4.0 | 2.7 | 1.9 | 16 |
| 197 | **Lentil and potato pie** | (2.7) | 2.3 | 0.7 | 0.9 | 0.7 | N | 0.07 | 0.5 | 0.6 | 0.8 | 0 |
| 198 | **Lentil roast** | (4.4) | (3.4) | 0.9 | 1.0 | 1.5 | N | 0.22 | 0.4 | 0.9 | 1.3 | 0 |
| 199 | *with egg* | (4.1) | (3.1) | 0.8 | 1.0 | 1.4 | N | 0.20 | 0.6 | 1.2 | 1.3 | 32 |
| 200 | **Lentil and nut roast** | (4.8) | (3.8) | 1.0 | 1.1 | 1.7 | N | 0.17 | 1.8 | 5.8 | 3.7 | 0 |
| 201 | *with egg* | (4.5) | (3.6) | 1.0 | 1.1 | 1.6 | N | 0.16 | 1.9 | 5.8 | 3.6 | 26 |
| 202 | **Lentil and rice roast** | (2.3) | 1.7 | 0.4 | 0.3 | 0.8 | N | N | 0.3 | 0.6 | 0.8 | 0 |
| 203 | *with egg* | (2.2) | 1.6 | 0.4 | 0.3 | 0.8 | N | N | 0.4 | 0.8 | 0.9 | 16 |
| 204 | **Mchicha** | (5.4) | 3.3 | 1.3 | 1.2 | 0.8 | 0.2 | 0.01 | 0.5 | 1.4 | 2.3 | 0 |
| 205 | **Moussaka**, vegetable | 1.6 | 1.2 | 0.4 | 0.6 | 0.3 | N | 0.02 | 2.8 | 3.2 | 2.8 | 26 |
| 206 | vegetable, retail | N | N | N | N | N | N | N | N | N | N | N |
| 207 | **Mushroom Dopiaza**, retail | 1.6 | 1.1 | 0.3 | 0.4 | 0.4 | 0.1 | 0.04 | 0.4 | 3.1 | 1.9 | 0 |
| 208 | **Nut croquettes**, *fried in sunflower oil* | 4.7 | (3.4) | (0.8) | (1.1) | (1.6) | N | 0.21 | 3.6 | 9.4 | 11.7 | 0 |
| 209 | *fried in vegetable oil* | 4.7 | (3.4) | (0.8) | (1.1) | (1.6) | N | 0.21 | 3.4 | 11.3 | 9.9 | 0 |
| 210 | **Nut cutlets**, retail, *fried in sunflower oil* | (2.7) | (1.7) | N | N | N | N | 0.23 | 2.5 | 8.1 | 10.5 | 0 |
| 211 | *fried in vegetable oil* | (2.7) | (1.7) | N | N | N | N | 0.23 | 2.4 | 9.7 | 9.0 | 0 |
| 212 | *grilled* | (2.9) | (1.8) | N | N | N | N | 0.25 | 1.4 | 6.5 | 4.5 | 0 |

# Vegetable Dishes

Inorganic constituents per 100g food

| No. 15- | Food | Na | K | Ca | Mg | P | Fe | Cu | Zn | Cl | Mn | Se | I |
|---|---|---|---|---|---|---|---|---|---|---|---|---|---|
| | | | | | | mg | | | | | | μg | |
| 195 | Lentil pie | 130 | 290 | 38 | 38 | 120 | 2.6 | 0.22 | 1.2 | 220 | N | (21) | N |
| 196 | Lentil and cheese pie | 170 | 270 | 140 | 29 | 190 | 2.4 | 0.21 | 1.3 | 270 | N | (18) | N |
| 197 | Lentil and potato pie | 310 | 300 | 15 | 25 | 79 | 1.8 | 0.18 | 0.8 | 500 | 0.2 | (11) | N |
| 198 | Lentil roast | 280 | 280 | 33 | 37 | 130 | 2.7 | 0.23 | 1.2 | 210 | N | (21) | N |
| 199 | with egg | 270 | 270 | 35 | 35 | 140 | 2.7 | 0.22 | 1.2 | 210 | N | (20) | N |
| 200 | Lentil and nut roast | 260 | 370 | 37 | 65 | 180 | 2.6 | 0.34 | 1.6 | 220 | 0.8 | (18) | N |
| 201 | with egg | 250 | 360 | 39 | 62 | 190 | 2.6 | 0.33 | 1.6 | 220 | 0.7 | (18) | N |
| 202 | Lentil and rice roast | 130 | 170 | 11 | 38 | 120 | 1.4 | 0.29 | 0.9 | 64 | (0.7) | (8) | N |
| 203 | with egg | 130 | 170 | 13 | 37 | 120 | 1.4 | 0.29 | 0.9 | 68 | (0.7) | (8) | 4 |
| 204 | Mchicha | 420 | 670 | 210 | 71 | 65 | 3.8 | 0.08 | 0.9 | 510 | 0.8 | (1) | N |
| 205 | Moussaka, vegetable | 170 | 240 | 80 | 15 | 91 | 0.7 | 0.12 | 0.6 | 290 | N | (3) | N |
| 206 | vegetable, retail | 450 | 330 | 76 | 25 | 100 | 1.0 | 0.06 | 0.5 | 690 | 0.2 | N | N |
| 207 | Mushroom Dopiaza, retail | N | 240 | 16 | 8 | 46 | 0.7 | 0.28 | 0.3 | N | 0.1 | 3 | N |
| 208 | Nut croquettes, fried in sunflower oil | 220 | 350 | 59 | 74 | 190 | 2.0 | 0.39 | 1.4 | 370 | 1.1 | (12) | 5 |
| 209 | fried in vegetable oil | 220 | 350 | 59 | 74 | 190 | 2.0 | 0.39 | 1.4 | 370 | 1.1 | (12) | 6 |
| 210 | Nut cutlets, retail, fried in sunflower oil | 210 | 190 | 84 | 60 | 140 | 1.3 | 0.32 | 1.0 | 340 | 0.5 | N | N |
| 211 | fried in vegetable oil | 210 | 190 | 84 | 60 | 140 | 1.3 | 0.32 | 1.0 | 340 | 0.5 | N | N |
| 212 | grilled | 220 | 210 | 89 | 64 | 150 | 1.4 | 0.34 | 1.1 | 360 | 0.5 | N | N |

| No. 15- | Food | Retinol µg | Carotene µg | Vitamin D µg | Vitamin E mg | Thiamin mg | Ribo-flavin mg | Niacin mg | Trypt 60 mg | Vitamin B6 mg | Vitamin B12 µg | Folate µg | Panto-thenate mg | Biotin µg | Vitamin C mg |
|---|---|---|---|---|---|---|---|---|---|---|---|---|---|---|---|
| 195 | **Lentil pie** | 0 | 1925 | 0 | (1.12) | 0.16 | 0.05 | 1.2 | 1.2 | 0.18 | 0 | 14 | N | N | 4 |
| 196 | **Lentil and cheese pie** | 54 | 46 | Tr | (0.87) | 0.13 | 0.20 | 1.0 | 2.0 | 0.17 | 0.2 | 16 | N | N | Tr |
| 197 | **Lentil and potato pie** | 18 | 23 | 0.2 | (0.07) | 0.14 | 0.03 | 0.5 | 0.8 | 0.24 | 0 | 12 | N | N | 2 |
| 198 | **Lentil roast** | 0 | 740 | 0 | (0.64) | 0.14 | 0.05 | 1.1 | 1.2 | 0.16 | 0 | 13 | N | N | 2 |
| 199 | *with egg* | 15 | 685 | 0.2 | (0.69) | 0.14 | 0.08 | 1.1 | 1.4 | 0.15 | 0.2 | 14 | N | N | 2 |
| 200 | **Lentil and nut roast** | 0 | 540 | 0 | (1.66) | 0.14 | 0.07 | 2.2 | 1.9 | 0.21 | 0 | 14 | N | N | Tr |
| 201 | *with egg* | 13 | 505 | 0.1 | (1.63) | 0.14 | 0.09 | 2.1 | 2.0 | 0.20 | 0.2 | 15 | N | N | Tr |
| 202 | **Lentil and rice roast** | 0 | 4 | 0 | (0.47) | 0.11 | 0.03 | 0.9 | 0.8 | N | 0 | 7 | N | N | Tr |
| 203 | *with egg* | 8 | 3 | 0.1 | (0.50) | 0.11 | 0.04 | 0.9 | 0.9 | N | 0.1 | 7 | N | N | Tr |
| 204 | **Mchicha** | 0 | 4115 | 0 | 3.01 | 0.09 | 0.09 | 1.4 | 1.0 | 0.19 | 0 | 87 | (0.29) | (0.4) | 17 |
| 205 | **Moussaka**, vegetable | 48 | 125 | 0.2 | 1.58 | 0.08 | 0.09 | 0.6 | 1.0 | 0.13 | 0.2 | 10 | 0.40 | N | 2 |
| 206 | vegetable, retail | N | N | N | N | 0.06 | 0.07 | 0.7 | 1.3 | 0.18 | Tr | 12 | N | N | N |
| 207 | **Mushroom Dopiaza**, retail | 0 | 285 | 0 | 1.56 | 0.08 | 0.09 | 1.3 | 0.3 | 0.13 | 0 | 12 | 0.63 | 3.8 | 3 |
| 208 | **Nut croquettes**, *fried in sunflower oil* | 0 | 6 | 0 | 7.83 | 0.16 | 0.12 | 4.1 | 1.9 | 0.17 | 0 | 15 | 0.81 | 23.6 | 1 |
| 209 | *fried in vegetable oil* | 0 | 6 | 0 | 4.67 | 0.16 | 0.12 | 4.1 | 1.9 | 0.17 | 0 | 15 | 0.75 | 19.5 | 1 |
| 210 | **Nut cutlets**, retail, *fried in sunflower oil* | 0 | 520 | 0 | 7.68 | 0.12 | 0.10 | 1.4 | 1.1 | N | 0 | 9 | (0.41) | (6.8) | 1 |
| 211 | *fried in vegetable oil* | 0 | 520 | 0 | 5.18 | 0.12 | 0.10 | 1.4 | 1.1 | N | 0 | 9 | (0.41) | (6.8) | 1 |
| 212 | *grilled* | 0 | 550 | 0 | 2.94 | 0.12 | 0.08 | 1.2 | 1.1 | N | 0 | 10 | (0.35) | (5.8) | 1 |

| No.<br>15- | Food | Description and main data sources |
|---|---|---|
| 213 | **Nut roast** | Recipe from review of recipe collection. Mixed nuts |
| 214 | *with egg* | Recipe from review of recipe collection. Mixed nuts |
| 215 | **Nut and rice roast** | Recipe from review of recipe collection. Mixed nuts |
| 216 | *with egg* | Recipe from review of recipe collection. Mixed nuts |
| 217 | **Nut and seed roast** | Recipe from review of recipe collection. Mixed nuts and sunflower seeds |
| 218 | *with egg* | Recipe from review of recipe collection. Mixed nuts and sunflower seeds |
| 219 | **Nut and vegetable roast** | Recipe from review of recipe collection. Mixed nuts |
| 220 | *with egg* | Recipe from review of recipe collection. Mixed nuts |
| 221 | **Nut, mushroom and rice roast** | Recipe adapted from a recipe from Leeds Polytechnic |
| 222 | **Okra with tomatoes and onion,** | |
| 223 | Greek | Recipe from a personal collection |
| 223 | West Indian | Recipe from review of recipe collection |
| 224 | **Pakora/bhajia,** aubergine, *fried*<br>*in vegetable oil* | Aubergine fried in chick pea flour batter. Recipe from review of recipe collection |
| 225 | cauliflower, *fried in vegetable*<br>*oil* | Cauliflower fried in chick pea flour batter. Recipe from review of recipe collection |
| 226 | onion, *fried in vegetable oil* | Onion fried in chick pea flour batter. Recipe from review of recipe collection |
| 227 | onion, retail | 14 samples purchased from take-away outlets |
| 228 | potato, *fried in vegetable oil* | Potato fried in chick pea flour batter. Recipe from review of recipe collection |
| 229 | potato and cauliflower, *fried*<br>*in vegetable oil* | Vegetables fried in chick pea flour batter. Recipe from review of recipe collection |

| No. | Food | Water | Total nitrogen | Protein | Fat | Carbohydrate | Energy value | |
|-----|------|-------|----------------|---------|-----|--------------|------|------|
| 15- | | g | g | g | g | g | kcal | kJ |
| 213 | **Nut roast** | 34.5 | 2.42 | 13.3 | 25.7 | 18.3 | 352 | 1467 |
| 214 | *with egg* | 37.3 | 2.41 | 13.3 | 24.7 | 16.9 | 338 | 1409 |
| 215 | **Nut and rice roast** | 33.5 | 2.17 | 12.0 | 22.6 | 25.2 | 346 | 1442 |
| 216 | *with egg* | 36.1 | 2.18 | 12.1 | 21.9 | 23.5 | 333 | 1391 |
| 217 | **Nut and seed roast** | 34.7 | 2.35 | 12.9 | 24.9 | 19.6 | 349 | 1454 |
| 218 | *with egg* | 37.5 | 2.35 | 13.0 | 23.9 | 18.1 | 335 | 1397 |
| 219 | **Nut and vegetable roast** | 40.1 | 2.02 | 11.1 | 21.9 | 19.6 | 314 | 1311 |
| 220 | *with egg* | 42.3 | 2.03 | 11.3 | 21.2 | 18.3 | 304 | 1269 |
| 221 | **Nut, mushroom and rice roast** | 48.3 | 1.43 | 7.8 | 15.3 | 24.3 | 260 | 1085 |
| 222 | **Okra with tomatoes and onion,** | | | | | | | |
| | Greek | 68.1 | 0.42 | 2.9 | 16.7 | 5.0 | 180 | 744 |
| 223 | West Indian | 75.6 | 0.45 | 3.1 | 7.7 | 4.7 | 97 | 405 |
| 224 | **Pakora/bhajia,** *aubergine, fried* | | | | | | | |
| | *in vegetable oil* | 45.2 | 1.10 | 6.9 | 22.5 | 17.2 | 295 | 1218 |
| 225 | cauliflower, *fried in vegetable* | | | | | | | |
| | *oil* | 40.1 | 1.43 | 8.9 | 24.1 | 18.2 | 322 | 1329 |
| 226 | onion, *fried in vegetable oil* | 40.6 | 1.57 | 9.8 | 14.7 | 26.2 | 271 | 1123 |
| 227 | onion, *retail* | 39.4 | 1.38 | 8.6 | 21.4 | 22.4 | 311 | 1296 |
| 228 | potato, *fried in vegetable oil* | 36.4 | 1.31 | 8.2 | 19.6 | 28.3 | 316 | 1310 |
| 229 | potato and cauliflower, *fried* | | | | | | | |
| | *in vegetable oil* | 41.9 | 1.27 | 7.8 | 22.1 | 20.3 | 306 | 1268 |

| No. 15- | Food | Starch | Total sugars | Glucose | Fructose | Galactose | Sucrose | Maltose | Lactose | Oligo-saccharides |
|---|---|---|---|---|---|---|---|---|---|---|
| 213 | **Nut roast** | 14.7 | 3.2 | N | N | 0 | N | N | 0 | 0.4 |
| 214 | *with egg* | 13.5 | 3.0 | N | N | 0 | N | N | 0 | 0.4 |
| 215 | **Nut and rice roast** | 22.0 | 2.9 | N | N | 0 | N | N | 0 | 0.3 |
| 216 | *with egg* | 20.5 | 2.7 | N | N | 0 | N | N | 0 | 0.3 |
| 217 | **Nut and seed roast** | 16.2 | 2.9 | N | N | 0 | 1.7 | N | 0 | 0.5 |
| 218 | *with egg* | 14.9 | 2.7 | N | N | 0 | 1.6 | N | 0 | 0.4 |
| 219 | **Nut and vegetable roast** | 15.4 | 3.9 | N | N | 0 | N | N | 0 | 0.4 |
| 220 | *with egg* | 14.3 | 3.6 | N | N | 0 | N | N | 0 | 0.3 |
| 221 | **Nut, mushroom and rice roast** | 21.2 | 2.8 | N | N | 0 | N | N | 0 | 0.3 |
| 222 | **Okra with tomatoes and onion,** Greek | 0.4 | 4.1 | 1.3 | 1.5 | 0 | 1.2 | 0 | 0 | 0.4 |
| 223 | West Indian | 0.6 | 3.9 | 1.2 | 1.5 | 0 | 1.1 | 0 | 0 | 0.3 |
| 224 | **Pakora/bhajia,** aubergine, *fried in vegetable oil* | 14.0 | 2.2 | 0.8 | 0.6 | 0 | 0.8 | 0 | 0 | 1.0 |
| 225 | cauliflower, *fried in vegetable oil* | 14.5 | 2.5 | 0.9 | 0.8 | 0 | 0.8 | 0 | 0 | 1.1 |
| 226 | onion, *fried in vegetable oil* | 20.5 | 3.3 | 0.8 | 0.7 | 0 | 1.8 | 0 | 0 | 2.4 |
| 227 | onion, retail | 14.6 | 7.8 | (1.7) | (1.7) | 0 | 4.5 | 0 | 0 | N |
| 228 | potato, *fried in vegetable oil* | 25.9 | 1.3 | 0.1 | 0.1 | 0 | 1.0 | 0 | 0 | 1.1 |
| 229 | potato and cauliflower, *fried in vegetable oil* | 17.1 | 1.9 | 0.5 | 0.4 | 0 | 1.0 | 0 | 0 | 1.2 |

**Fibre fractions, phytic acid and fatty acids, g per 100g food**
**Cholesterol, mg per 100g food**

| No. 15- | Food | Dietary fibre, g | | Fibre fractions, g | | | | Phytic acid g | Fatty acids, g | | | Cholesterol mg |
| | | Southgate method | Englyst method | Cellulose | Non-cellulosic polysaccharide | | Lignin | | Satd | Mono-unsatd | Poly-unsatd | |
| | | | | | Soluble | Insoluble | | | | | | | |
| 213 | **Nut roast** | 5.4 | (4.2) | 1.0 | 1.3 | 1.9 | N | 0.19 | 3.9 | 12.8 | 7.5 | 0 |
| 214 | *with egg* | 5.0 | (3.9) | 0.9 | 1.2 | 1.8 | N | 0.17 | 3.9 | 12.2 | 7.0 | 33 |
| 215 | **Nut and rice roast** | 5.0 | (3.7) | 0.9 | 1.0 | 1.7 | N | N | 3.5 | 11.2 | 6.6 | 0 |
| 216 | *with egg* | 4.7 | (3.5) | 0.9 | 1.0 | 1.6 | N | N | 3.5 | 10.8 | 6.3 | 29 |
| 217 | **Nut and seed roast** | (5.3) | (4.2) | 1.0 | 1.3 | 1.9 | N | 0.55 | 3.4 | 10.6 | 9.5 | 0 |
| 218 | *with egg* | (4.9) | (3.9) | 0.9 | 1.2 | 1.8 | N | 0.50 | 3.4 | 10.2 | 8.9 | 33 |
| 219 | **Nut and vegetable roast** | 5.0 | (3.9) | 1.0 | 1.3 | 1.7 | N | 0.17 | 3.2 | 10.6 | 6.8 | 0 |
| 220 | *with egg* | 4.7 | (3.6) | 0.9 | 1.2 | 1.5 | N | 0.16 | 3.3 | 10.2 | 6.4 | 29 |
| 221 | **Nut, mushroom and rice roast** | 3.9 | (2.7) | 0.7 | 0.8 | 1.3 | N | N | 2.2 | 7.0 | 5.2 | 0 |
| 222 | **Okra with tomatoes and onion,** | | | | | | | | | | | |
| | Greek | 4.6 | 4.1 | 1.3 | 2.4 | 0.4 | 0.5 | N | 2.5 | 11.1 | 2.1 | 0 |
| 223 | West Indian | 4.7 | 4.2 | 1.3 | 2.4 | 0.5 | 0.6 | N | 1.0 | 2.5 | 3.5 | 0 |
| 224 | **Pakora/bhajia,** aubergine, *fried in vegetable oil* | (5.8) | (4.8) | (1.3) | (1.7) | (1.8) | (0.7) | 0.25 | 2.4 | 7.7 | 10.8 | 0 |
| 225 | cauliflower, *fried in vegetable oil* | (5.7) | (4.7) | (1.1) | (1.7) | (1.9) | (0.6) | 0.30 | 2.6 | 8.1 | 11.7 | 0 |
| 226 | onion, *fried in vegetable oil* | (6.9) | (5.5) | (1.4) | (1.8) | (2.4) | (0.8) | 0.38 | 1.5 | 4.8 | 7.1 | 0 |
| 227 | onion, *retail* | (6.9) | (5.5) | (1.4) | (1.8) | (2.4) | (0.8) | (0.38) | 1.9 | 11.8 | 6.6 | 0 |
| 228 | potato, *fried in vegetable oil* | (5.7) | (4.5) | (1.1) | (1.6) | (1.8) | (0.6) | 0.27 | 2.0 | 6.6 | 9.4 | 0 |
| 229 | potato and cauliflower, *fried in vegetable oil* | (4.8) | (3.9) | (1.0) | (1.4) | (1.6) | (0.5) | 0.25 | 2.5 | 7.4 | 10.7 | 0 |

| No. 15- | Food | Na | K | Ca | Mg | P | Fe | Cu | Zn | Cl | Mn | Se | I |
|---|---|---|---|---|---|---|---|---|---|---|---|---|---|
| | | | | | | mg | | | | | | μg | |
| 213 | **Nut roast** | 340 | 470 | 61 | 110 | 270 | 1.9 | 0.42 | 1.9 | 550 | 1.4 | (11) | 7 |
| 214 | *with egg* | 330 | 450 | 61 | 100 | 270 | 1.9 | 0.40 | 1.9 | 520 | 1.3 | (11) | 11 |
| 215 | **Nut and rice roast** | 440 | 440 | 48 | 110 | 280 | 1.7 | 0.49 | 1.9 | 730 | 1.5 | (9) | (6) |
| 216 | *with egg* | 420 | 420 | 49 | 100 | 270 | 1.7 | 0.46 | 1.8 | 690 | 1.4 | (9) | (10) |
| 217 | **Nut and seed roast** | 310 | 460 | 65 | 130 | 300 | 2.4 | 0.60 | 2.1 | 490 | 1.4 | (17) | 6 |
| 218 | *with egg* | 300 | 440 | 65 | 120 | 290 | 2.4 | 0.56 | 2.1 | 470 | 1.3 | (16) | 10 |
| 219 | **Nut and vegetable roast** | 300 | 450 | 60 | 89 | 230 | 1.7 | 0.36 | 1.6 | 490 | 1.2 | (10) | 6 |
| 220 | *with egg* | 290 | 430 | 60 | 84 | 230 | 1.8 | 0.34 | 1.6 | 470 | 1.1 | (10) | 10 |
| 221 | **Nut, mushroom and rice roast** | 290 | 340 | 32 | 72 | 200 | 1.3 | 0.44 | 1.3 | 350 | 1.1 | (7) | (4) |
| 222 | **Okra with tomatoes and onion,** Greek | 130 | 390 | 150 | 66 | 64 | 1.2 | 0.13 | 0.6 | 240 | N | (1) | N |
| 223 | West Indian | 530 | 410 | 160 | 73 | 68 | 1.4 | 0.13 | 0.6 | 850 | N | (1) | N |
| 224 | **Pakora/bhajia**, aubergine, *fried in vegetable oil* | 270 | 470 | 65 | 48 | 120 | 2.9 | 0.21 | 1.1 | 430 | 0.7 | (1) | N |
| 225 | cauliflower, *fried in vegetable oil* | 290 | 590 | 74 | 53 | 160 | 3.2 | 0.22 | 1.4 | 450 | 0.9 | 1 | N |
| 226 | onion, *fried in vegetable oil* | 220 | 550 | 98 | 61 | 170 | 4.2 | 0.32 | 1.6 | 350 | 1.0 | 1 | (3) |
| 227 | onion, *retail* | 360 | (550) | (98) | (61) | (170) | 3.2 | (0.32) | (1.5) | 560 | (1.0) | (1) | (3) |
| 228 | potato, *fried in vegetable oil* | 390 | 590 | 67 | 56 | 140 | 3.2 | 0.27 | 1.3 | 630 | 0.8 | 1 | N |
| 229 | potato and cauliflower, *fried in vegetable oil* | 590 | 480 | 68 | 61 | 140 | 3.1 | 0.28 | 1.2 | 930 | 0.8 | 1 | N |

| No. 15- | Food | Retinol µg | Carotene µg | Vitamin D µg | Vitamin E mg | Thiamin mg | Ribo-flavin mg | Niacin mg | Trypt/60 mg | Vitamin B6 mg | Vitamin B12 µg | Folate µg | Panto-thenate mg | Biotin µg | Vitamin C mg |
|---|---|---|---|---|---|---|---|---|---|---|---|---|---|---|---|
| 213 | **Nut roast** | 0 | 15 | 0 | 3.33 | 0.20 | 0.24 | 5.1 | 2.8 | 0.25 | 0 | 26 | 0.63 | 30.6 | Tr |
| 214 | *with egg* | 16 | 14 | 0.2 | 3.17 | 0.19 | 0.26 | 4.7 | 2.9 | 0.24 | 0.2 | 26 | 0.70 | 29.7 | Tr |
| 215 | **Nut and rice roast** | 0 | 13 | 0 | 3.00 | 0.21 | 0.21 | 4.7 | 2.6 | N | 0 | 24 | N | N | Tr |
| 216 | *with egg* | 14 | 12 | 0.1 | 2.88 | 0.20 | 0.23 | 4.4 | 2.7 | N | 0.2 | 24 | N | N | Tr |
| 217 | **Nut and seed roast** | 0 | 17 | 0 | 7.16 | 0.34 | 0.24 | 4.5 | 2.8 | (0.20) | 0 | (23) | (0.49) | (22.2) | Tr |
| 218 | *with egg* | 16 | 16 | 0.2 | 6.72 | 0.32 | 0.25 | 4.2 | 3.0 | (0.19) | 0.2 | (23) | (0.57) | (21.9) | Tr |
| 219 | **Nut and vegetable roast** | 0 | 850 | 0 | 3.29 | 0.19 | 0.21 | 4.3 | 2.4 | 0.23 | 0 | 24 | 0.57 | 23.9 | 2 |
| 220 | *with egg* | 14 | 790 | 0.1 | 3.16 | 0.19 | 0.22 | 4.0 | 2.5 | 0.22 | 0.2 | 24 | 0.64 | 23.5 | 1 |
| 221 | **Nut, mushroom and rice roast** | 0 | 805 | 0 | 2.50 | 0.17 | 0.18 | 3.4 | 1.7 | N | Tr | 20 | N | N | 1 |
| 222 | **Okra with tomatoes and onion,** Greek | 0 | 635 | 0 | N | 0.18 | 0.04 | 1.0 | 0.4 | 0.21 | 0 | 42 | 0.25 | N | 12 |
| 223 | West Indian | 0 | 750 | 0 | N | 0.18 | 0.05 | 1.1 | 0.5 | 0.21 | 0 | 44 | 0.25 | 0.4 | 13 |
| 224 | **Pakora/bhajia,** aubergine, *fried in vegetable oil* | 0 | 150 | 0 | 5.85 | (0.11) | (0.09) | (0.7) | 0.9 | (0.17) | 0 | (31) | (0.56) | N | 2 |
| 225 | cauliflower, *fried in vegetable oil* | 0 | 140 | 0 | 6.31 | (0.19) | (0.12) | (1.1) | 1.5 | (0.27) | 0 | (46) | (0.92) | N | 20 |
| 226 | onion, *fried in vegetable oil* | 0 | 110 | 0 | 4.36 | (0.19) | (0.12) | (1.2) | 1.3 | (0.24) | 0 | (40) | (0.78) | (0.3) | 2 |
| 227 | onion, retail | 27 | 56 | Tr | 2.70 | (0.10) | (0.08) | (1.2) | 1.4 | (0.24) | 0 | (40) | (0.78) | (0.3) | (2) |
| 228 | potato, *fried in vegetable oil* | 0 | 135 | 0 | 5.25 | (0.22) | (0.10) | (1.0) | 1.2 | (0.35) | 0 | (38) | (0.79) | N | 5 |
| 229 | potato and cauliflower, *fried in vegetable oil* | 0 | 28 | 0 | 5.54 | (0.17) | (0.09) | (0.9) | 1.3 | (0.24) | 0 | (33) | (0.66) | N | 7 |

No. Food
**15-**

Description and main data sources

| No. | Food | Description and main data sources |
|---|---|---|
| 230 | **Pakora/bhajia**, potato, carrot and pea, *fried in vegetable oil* | Vegetables fried in chick pea flour batter. Recipe from review of recipe collection |
| 231 | spinach, *fried in vegetable oil* | Spinach fried in chick pea flour batter. Recipe from review of rceipe collection |
| 232 | vegetable, retail | Recipe from manufacturer |
| 233 | **Pancakes**, stuffed with vegetables | Tomato, mushroom and onion stuffing. Recipe adapted from a recipe from Leeds Polytechnic |
| 234 | stuffed with vegetables, wholemeal | Tomato, mushroom and onion stuffing. Recipe adapted from a recipe from Leeds Polytechnic |
| 235 | **Pastichio** | Greek dish. Macaroni, lentil and vegetables with white sauce. Recipe from a personal collection |
| 236 | **Pasty**, vegetable | Recipe from dietary survey records |
| 237 | vegetable, wholemeal | Recipe from dietary survey records |
| 238 | **Peppers**, stuffed with rice | Green peppers with rice, tomato and raisin stuffing. Recipe from review of recipe collection |
| 239 | stuffed with vegetables, cheese topping | Green peppers with mixed vegetable stuffing. Recipe from a review of recipe collection |
| 240 | **Pesto sauce** | Italian dish. Basil and parmesan sauce for pasta. Recipe from Waltham Forest College |
| 241 | **Pie**, Quorn and vegetable | Recipe from Marlow Foods Ltd |
| 242 | spinach | Greek dish. Spinach, leek, feta cheese and egg pie. Recipe from a personal collection |
| 243 | vegetable | Recipe from review of recipe collection |
| 244 | vegetable, wholemeal | Recipe from review of recipe collection |

| No. 15- | Food | Water g | Total nitrogen g | Protein g | Fat g | Carbohydrate g | Energy value kcal | kJ |
|---|---|---|---|---|---|---|---|---|
| 230 | **Pakora/bhajia**, potato, carrot and pea, *fried in vegetable oil* | 28.5 | 1.73 | 10.9 | 22.6 | 28.8 | 357 | 1477 |
| 231 | spinach, *fried in vegetable oil* | 22.3 | 2.02 | 12.6 | 21.9 | 29.8 | 361 | 1493 |
| 232 | vegetable, retail | 50.5 | 1.04 | 6.4 | 14.7 | 21.4 | 235 | 975 |
| 233 | **Pancakes**, stuffed with vegetables | 70.7 | 0.70 | 4.1 | 8.4 | 15.4 | 149 | 625 |
| 234 | stuffed with vegetables, wholemeal | 70.7 | 0.78 | 4.6 | 8.5 | 13.4 | 145 | 606 |
| 235 | **Pastichio** | 69.6 | 1.07 | 6.5 | 7.0 | 14.1 | 141 | 593 |
| 236 | **Pasty**, vegetable | 47.0 | 0.70 | 4.1 | 14.9 | 33.3 | 274 | 1150 |
| 237 | vegetable, wholemeal | 47.1 | 0.90 | 5.3 | 15.2 | 28.0 | 262 | 1098 |
| 238 | **Peppers**, stuffed with rice | 78.0 | 0.25 | 1.5 | 2.4 | 15.4 | 85 | 360 |
| 239 | stuffed with vegetables, cheese topping | 78.2 | 0.57 | 3.4 | 6.7 | 9.8 | 111 | 463 |
| 240 | **Pesto sauce** | 23.2 | 3.28 | 20.4 | 47.5 | 2.0 | 517 | 2139 |
| 241 | **Pie**, Quorn and vegetable | 64.5 | 1.11 | 6.9 | 11.5 | 14.7 | 186 | 776 |
| 242 | spinach | 66.4 | 1.07 | 6.7 | 13.7 | 8.1 | 180 | 748 |
| 243 | vegetable | 68.7 | 0.52 | 3.0 | 7.6 | 18.9 | 151 | 634 |
| 244 | vegetable, wholemeal | 68.7 | 0.62 | 3.6 | 7.8 | 16.5 | 146 | 611 |

# Vegetable Dishes

## Carbohydrate fractions, g per 100g food

| No. Food 15- | Starch | Total sugars | Individual sugars | | | | | | Oligo- saccharides |
|---|---|---|---|---|---|---|---|---|---|
| | | | Glucose | Fructose | Galactose | Sucrose | Maltose | Lactose | |
| 230 **Pakora/bhajia**, potato, carrot and pea, *fried in vegetable oil* | 24.9 | 2.0 | 0.2 | 0.2 | 0 | 1.6 | 0 | 0 | 1.8 |
| 231 spinach, *fried in vegetable oil* | 25.9 | 2.0 | 0.2 | 0.3 | 0 | 1.6 | 0 | 0 | 1.9 |
| 232 vegetable, retail | 17.5 | 2.4 | 0.7 | 0.6 | 0 | 1.1 | Tr | 0 | 1.3 |
| 233 **Pancakes**, stuffed with vegetables | 11.0 | 4.0 | 0.8 | 0.8 | 0 | 0.3 | Tr | 2.0 | 0.4 |
| 234 stuffed with vegetables, wholemeal | 8.9 | 4.1 | 0.9 | 0.8 | 0 | 0.4 | Tr | 2.0 | 0.4 |
| 235 **Pastichio** | 11.1 | 2.6 | 0.5 | 0.5 | 0 | 0.4 | 0 | 1.1 | 0.5 |
| 236 **Pasty**, vegetable | 31.4 | 1.8 | 0.5 | 0.4 | 0 | 0.6 | 0.1 | 0.2 | 0.1 |
| 237 vegetable, wholemeal | 25.9 | 2.0 | 0.5 | 0.4 | 0 | 0.9 | Tr | 0.2 | 0.1 |
| 238 **Peppers**, stuffed with rice | 9.3 | 5.8 | 2.7 | 2.9 | 0 | 0.1 | 0 | 0 | 0.2 |
| 239 stuffed with vegetables, cheese topping | 4.8 | 4.6 | 1.2 | 1.3 | 0 | 2.1 | 0 | 0 | 0.4 |
| 240 **Pesto sauce** | 0.5 | 1.5 | N | N | 0 | N | 0 | Tr | Tr |
| 241 **Pie**, Quorn and vegetable | 10.6 | 3.7 | 0.5 | 0.4 | 0 | 1.0 | 0.1 | 1.7 | 0.5 |
| 242 spinach | 5.8 | 2.2 | 0.5 | 0.6 | 0 | 0.4 | Tr | 0.6 | 0.1 |
| 243 vegetable | 15.7 | 3.0 | 1.1 | 1.1 | 0 | 0.6 | Tr | 0.1 | 0.3 |
| 244 vegetable, wholemeal | 13.2 | 3.1 | 1.1 | 1.1 | 0 | 0.7 | Tr | 0.1 | 0.3 |

Fibre fractions, phytic acid and fatty acids, g per 100g food
Cholesterol, mg per 100g food

| No. 15- | Food | Dietary fibre, g | | Fibre fractions, g | | | | Phytic acid | Fatty acids, g | | | Cholesterol |
|---|---|---|---|---|---|---|---|---|---|---|---|---|
| | | Southgate method | Englyst method | Cellulose | Non-cellulosic polysaccharide | | Lignin | acid g | Satd | Mono-unsatd | Poly-unsatd | mg |
| | | | | | Soluble | Insoluble | | | | | | | |
| 230 | **Pakora/bhajia**, potato, carrot and pea, *fried in vegetable oil* | (7.7) | (6.1) | (1.5) | (2.0) | (2.6) | (1.0) | 0.41 | 2.3 | 7.6 | 10.9 | 0 |
| 231 | spinach, *fried in vegetable oil* | (9.2) | (7.0) | (1.8) | (2.2) | (3.0) | (1.1) | 0.47 | 2.2 | 7.2 | 10.6 | 0 |
| 232 | vegetable, retail | (4.5) | (3.6) | (0.9) | (1.3) | (1.4) | (0.4) | 0.19 | 1.0 | 7.6 | 4.8 | 0 |
| 233 | **Pancakes**, stuffed with vegetables | 1.2 | 1.0 | 0.2 | 0.5 | 0.3 | Tr | 0.03 | 2.7 | 2.6 | 2.5 | 30 |
| 234 | stuffed with vegetables, wholemeal | 1.9 | 1.8 | 0.4 | 0.5 | 0.9 | Tr | 0.08 | 2.7 | 2.6 | 2.6 | 30 |
| 235 | **Pastichio** | 2.1 | 1.2 | 0.3 | 0.4 | 0.5 | 0.1 | 0.04 | 3.6 | 2.0 | 0.7 | 36 |
| 236 | **Pasty**, vegetable | 2.2 | 1.9 | 0.2 | 1.0 | 0.7 | Tr | 0.05 | 3.7 | 4.6 | 5.8 | 1 |
| 237 | vegetable, wholemeal | 4.1 | 4.1 | 0.7 | 1.2 | 2.2 | 0.1 | 0.19 | 3.7 | 4.6 | 5.9 | 1 |
| 238 | **Peppers**, stuffed with rice | (2.0) | 1.3 | 0.5 | 0.6 | 0.3 | 0.2 | 0.06 | 0.4 | 0.7 | 1.2 | 0 |
| 239 | stuffed with vegetables, cheese topping | (2.4) | 1.5 | 0.5 | 0.5 | 0.5 | 0.2 | 0.03 | 2.0 | 2.1 | 2.2 | 7 |
| 240 | **Pesto sauce** | N | N | N | N | N | N | 0.01 | (12.8) | (22.3) | (10.0) | 43 |
| 241 | **Pie**, Quorn and vegetable | (2.2) | 2.0 | N | N | N | Tr | (0.02) | 3.7 | 3.6 | 3.4 | 20 |
| 242 | spinach | (3.0) | 1.9 | 0.7 | 0.8 | 0.4 | 0.1 | 0.01 | (4.5) | (5.3) | (2.7) | 47 |
| 243 | vegetable | (1.8) | 1.5 | 0.4 | 0.7 | 0.4 | Tr | 0.03 | 1.9 | 2.3 | 3.0 | 0 |
| 244 | vegetable, wholemeal | (2.7) | 2.5 | 0.6 | 0.8 | 1.1 | 0.1 | 0.10 | 1.9 | 2.3 | 3.0 | 0 |

| No. 15- | Food | Na | K | Ca | Mg | P | mg Fe | Cu | Zn | Cl | Mn | µg Se | I |
|---------|------|----|----|----|----|----|----|----|----|----|----|----|----|
| 230 | **Pakora/bhajia**, potato, carrot and pea, *fried in vegetable oil* | 260 | 590 | 98 | 67 | 190 | 4.5 | 0.34 | 1.7 | 400 | 1.1 | 1 | N |
| 231 | spinach, *fried in vegetable oil* | 820 | 770 | 160 | 94 | 220 | 5.7 | 0.39 | 2.1 | 1230 | 1.4 | (1) | N |
| 232 | vegetable, retail | 61 | 490 | 99 | 47 | 130 | 3.7 | 0.21 | 1.1 | 71 | 0.7 | (1) | N |
| 233 | **Pancakes**, stuffed with vegetables | 150 | 230 | 80 | 13 | 83 | 0.8 | 0.12 | 0.4 | 260 | 0.2 | 3 | (14) |
| 234 | stuffed with vegetables, wholemeal | 150 | 250 | 65 | 28 | 110 | 1.1 | 0.17 | 0.8 | 260 | 0.5 | 10 | N |
| 235 | **Pastichio** | 260 | 230 | 76 | 22 | 110 | 1.4 | 0.16 | 0.8 | 420 | (0.1) | (4) | (11) |
| 236 | **Pasty**, vegetable | 320 | 140 | 61 | 13 | 56 | 1.0 | 0.08 | 0.3 | 510 | 0.3 | 2 | 9 |
| 237 | vegetable, wholemeal | 310 | 210 | 23 | 50 | 130 | 1.7 | 0.19 | 1.2 | 500 | 1.2 | 20 | N |
| 238 | **Peppers**, stuffed with rice | 180 | 190 | 16 | 13 | 36 | 0.6 | 0.08 | 0.3 | 290 | 0.2 | (2) | (3) |
| 239 | stuffed with vegetables, cheese topping | 340 | 230 | 62 | 16 | 79 | 0.6 | 0.11 | 0.4 | 540 | 0.1 | 2 | N |
| 240 | **Pesto sauce** | 480 | 250 | 560 | 69 | 480 | 2.4 | 0.38 | 3.6 | 810 | 1.4 | N | N |
| 241 | **Pie**, Quorn and vegetable | 350 | (110) | (91) | (10) | (79) | (0.4) | (0.03) | (0.4) | (460) | (0.1) | (2) | N |
| 242 | spinach | 420 | 390 | 180 | 36 | 120 | 1.8 | 0.06 | 0.8 | 610 | 0.4 | N | N |
| 243 | vegetable | 210 | 250 | 41 | 14 | 58 | 0.8 | 0.10 | 0.3 | 360 | 0.2 | (2) | (6) |
| 244 | vegetable, wholemeal | 210 | 280 | 23 | 32 | 95 | 1.1 | 0.15 | 0.7 | 350 | 0.6 | (10) | N |

# Vegetable Dishes

| No. 15- | Food | Retinol μg | Carotene μg | Vitamin D μg | Vitamin E mg | Thiamin mg | Riboflavin mg | Niacin mg | Trypt/60 mg | Vitamin B6 mg | Vitamin B12 μg | Folate μg | Pantothenate mg | Biotin μg | Vitamin C mg |
|---|---|---|---|---|---|---|---|---|---|---|---|---|---|---|---|
| 230 | **Pakora/bhajia**, potato, carrot and pea, *fried in vegetable oil* | 0 | 640 | 0 | 6.29 | (0.20) | (0.13) | (1.2) | 1.5 | (0.25) | 0 | (45) | (0.89) | N | 2 |
| 231 | spinach, *fried in vegetable oil* | 0 | 1355 | 0 | 6.64 | (0.20) | (0.18) | N | 1.8 | N | 0 | (68) | N | N | 6 |
| 232 | vegetable, retail | 0 | 965 | 0 | 3.66 | (0.17) | (0.10) | (1.2) | 1.0 | (0.23) | 0 | (30) | (0.53) | N | 7 |
| 233 | **Pancakes**, stuffed with vegetables | 91 | 150 | 0.7 | 0.62 | 0.09 | 0.11 | 0.8 | 0.9 | 0.11 | 0.3 | 11 | 0.45 | 3.5 | 3 |
| 234 | stuffed with vegetables, wholemeal | 91 | 150 | 0.7 | 0.78 | 0.12 | 0.11 | 1.3 | 1.0 | 0.15 | 0.3 | 14 | 0.51 | 4.4 | 3 |
| 235 | **Pastichio** | 71 | 115 | 0.3 | 0.63 | 0.07 | 0.10 | 0.5 | 1.2 | 0.11 | 0.3 | 5 | 0.35 | (2.0) | 3 |
| 236 | **Pasty**, vegetable | 140 | 620 | 1.4 | 0.24 | 0.13 | 0.01 | 0.7 | 0.8 | 0.10 | 0 | 9 | 0.17 | 0.5 | 2 |
| 237 | vegetable, wholemeal | 140 | 620 | 1.4 | 0.65 | 0.17 | 0.03 | 2.1 | 1.0 | 0.20 | 0 | 15 | 0.31 | 2.7 | 2 |
| 238 | **Peppers**, stuffed with rice | 0 | 275 | 0 | 1.07 | 0.03 | 0.01 | 0.5 | 0.3 | 0.18 | 0 | 13 | (0.10) | N | 34 |
| 239 | stuffed with vegetables, cheese topping | 23 | 300 | Tr | 1.76 | 0.04 | 0.06 | 0.7 | 0.7 | 0.17 | 0.1 | 11 | 0.27 | N | 27 |
| 240 | **Pesto sauce** | (150) | (650) | (0.1) | 3.78 | 0.16 | 0.27 | 0.9 | 4.6 | N | 0.8 | N | N | N | 4 |
| 241 | **Pie**, Quorn and vegetable | 105 | 690 | 0.8 | 0.39 | N | 0.10 | 0.4 | (0.9) | 0.06 | 0.3 | 5 | 0.21 | (2.9) | 1 |
| 242 | spinach | 105 | 2105 | 0.8 | 1.51 | 0.12 | 0.13 | 0.8 | 1.6 | 0.20 | 0.4 | 52 | (0.35) | (2.3) | 9 |
| 243 | vegetable | 71 | 1270 | 0.7 | 0.46 | 0.12 | 0.10 | 1.2 | 0.6 | 0.15 | 0 | 15 | 0.24 | (1.0) | 9 |
| 244 | vegetable, wholemeal | 71 | 1270 | 0.7 | 0.66 | 0.14 | 0.11 | 1.8 | 0.7 | 0.20 | 0 | 18 | 0.30 | (2.1) | 9 |

No. Food
**15-**

Description and main data sources

| No. | Food | Description and main data sources |
|-----|------|-----------------------------------|
| 245 | **Pilaf**, rice with spinach | Greek dish. Recipe from a personal collection |
| 246 | rice with tomato | Greek dish. Recipe from a personal collection |
| 247 | **Pilau**, egg and potato | Rice-based dish. Recipe from review of recipe collection |
| 248 | egg and potato, brown rice | Rice-based dish. Recipe from review of recipe collection |
| 249 | mushroom | Rice-based dish. Recipe from a personal collection |
| 250 | plain | Rice-based dish. Recipe from a personal collection |
| 251 | vegetable | Rice-based dish with frozen mixed vegetables. Recipe from a personal collection |
| 252 | **Pizza**, cheese and tomato | Recipe from Holland et al, 1991b |
| 253 | cheese and tomato, wholemeal | Recipe adapted from No 252 |
| 254 | cheese and tomato, retail, frozen | |
| 255 | tomato | 10 samples, 2 brands |
| 256 | tomato, wholemeal | Recipe from review of recipe collection |
| 257 | **Potato cakes**, *fried in lard* | Recipe from review of recipe collection |
| 258 | *fried in vegetable oil* | Recipe from review of recipe collection |
| 259 | **Potato, leek and celery bake** | Recipe from review of recipe collection |
| 260 | **Potatoes**, duchesse | Recipe from dietary survey records |
| 261 | **Potatoes with eggs** | Baked potato, egg and butter shapes. Recipe from review of recipe collection |
| 262 | **Quorn korma** | Greek dish. Fried potatoes with onion and egg. Recipe from a personal collection |
| 263 | **Ratatouille** | Quorn in korma curry sauce. Recipe from manufacturer |
| 264 | retail | Recipe adapted from Wiles et al, 1980 |
| | | 9 frozen samples, 3 brands; shallow fried then simmered for 35–40 minutes |

| No. 15- | Food | Water g | Total nitrogen g | Protein g | Fat g | Carbohydrate g | Energy value kcal | Energy value kJ |
|---|---|---|---|---|---|---|---|---|
| 245 | **Pilaf**, rice with spinach | 73.2 | 0.58 | 3.5 | 2.5 | 17.7 | 103 | 436 |
| 246 | rice with tomato | 67.1 | 0.43 | 2.5 | 3.3 | 28.0 | 144 | 612 |
| 247 | **Pilau**, egg and potato | 70.0 | 0.65 | 4.0 | 3.9 | 20.4 | 127 | 538 |
| 248 | egg and potato, brown rice | 69.0 | 0.65 | 4.0 | 3.8 | 21.1 | 129 | 543 |
| 249 | mushroom | 69.9 | 0.44 | 2.4 | 4.4 | 23.9 | 138 | 582 |
| 250 | plain | 69.4 | 0.38 | 2.3 | 4.6 | 24.8 | 141 | 598 |
| 251 | vegetable | 69.3 | 0.43 | 2.6 | 4.4 | 24.1 | 138 | 585 |
| 252 | **Pizza**, cheese and tomato | 51.0 | 1.49 | 9.1 | 11.8 | 25.2 | 237 | 995 |
| 253 | cheese and tomato, wholemeal | 51.0 | 1.65 | 10.0 | 12.1 | 21.0 | 228 | 955 |
| 254 | cheese and tomato, retail, frozen | 49.3 | 1.31 | 7.5 | 10.7 | 32.9 | 250 | 1050 |
| 255 | tomato | 60.8 | 0.57 | 3.3 | 10.6 | 22.6 | 193 | 808 |
| 256 | tomato, wholemeal | 60.8 | 0.71 | 4.2 | 10.8 | 19.1 | 185 | 775 |
| 257 | **Potato cakes**, *fried in lard* | 55.8 | 0.66 | 3.9 | 8.5 | 31.4 | 209 | 881 |
| 258 | *fried in vegetable oil* | 55.8 | 0.66 | 3.9 | 8.5 | 31.4 | 210 | 883 |
| 259 | **Potato, leek and celery bake** | 73.1 | 0.69 | 4.4 | 6.5 | 12.6 | 122 | 512 |
| 260 | **Potatoes**, duchesse | 73.6 | 0.49 | 3.1 | 5.3 | 17.0 | 123 | 518 |
| 261 | **Potatoes with eggs** | 43.7 | 1.26 | 7.8 | 23.8 | 19.1 | 316 | 1318 |
| 262 | **Quorn korma** | 71.1 | 0.62 | 3.7 | 7.0 | 16.7 | 140 | 590 |
| 263 | **Ratatouille** | 83.9 | 0.21 | 1.3 | 7.0 | 3.8 | 82 | 340 |
| 264 | retail | 85.5 | 0.20 | 1.2 | 6.6 | 3.7 | 78 | 324 |

## Carbohydrate fractions, g per 100g food

| No. 15- | Food | Starch | Total sugars | Individual sugars | | | | | | Oligo-saccharides |
|---|---|---|---|---|---|---|---|---|---|---|
| | | | | Glucose | Fructose | Galactose | Sucrose | Maltose | Lactose | |
| 245 | **Pilaf**, rice with spinach | 16.0 | 1.5 | 0.5 | 0.5 | 0 | 0.5 | 0 | 0 | 0.2 |
| 246 | rice with tomato | 27.0 | 0.9 | 0.4 | 0.4 | 0 | 0.1 | 0 | 0 | 0.1 |
| 247 | **Pilau**, egg and potato | 18.6 | 1.4 | 0.5 | 0.4 | 0 | 0.5 | 0 | 0 | 0.5 |
| 248 | egg and potato, brown rice | 18.9 | 1.6 | 0.6 | 0.4 | 0 | 0.6 | 0 | 0 | 0.5 |
| 249 | mushroom | 22.9 | 0.7 | 0.3 | 0.2 | 0 | 0.2 | 0 | Tr | 0.3 |
| 250 | plain | 23.8 | 0.7 | 0.3 | 0.2 | 0 | 0.2 | 0 | Tr | 0.3 |
| 251 | vegetable | 22.7 | 1.1 | 0.3 | 0.2 | 0 | 0.6 | 0 | Tr | 0.3 |
| 252 | **Pizza**, cheese and tomato | 23.0 | 2.2 | 0.6 | 0.6 | 0 | 0.9 | 0.1 | Tr | 0 |
| 253 | cheese and tomato, wholemeal | 18.7 | 2.4 | 0.6 | 0.6 | 0 | 1.1 | Tr | Tr | 0 |
| 254 | cheese and tomato, retail, frozen | 26.0 | 6.9 | 0.6 | 1.0 | 0 | 0.3 | 4.8 | 0.2 | 0 |
| 255 | tomato | 19.6 | 3.0 | 1.0 | 1.1 | 0 | 0.8 | Tr | 0 | 0 |
| 256 | tomato, wholemeal | 16.0 | 3.2 | 1.0 | 1.1 | 0 | 1.0 | Tr | 0 | 0 |
| 257 | **Potato cakes**, *fried in lard* | 30.1 | 1.3 | 0.3 | 0.2 | 0 | 0.3 | 0 | 0.4 | 0 |
| 258 | *fried in vegetable oil* | 30.1 | 1.3 | 0.3 | 0.2 | 0 | 0.3 | 0 | 0.4 | 0 |
| 259 | **Potato, leek and celery bake** | 10.8 | 1.6 | 0.2 | 0.1 | 0 | 0.2 | Tr | 1.1 | Tr |
| 260 | **Potatoes**, duchesse | 16.3 | 0.7 | 0.2 | 0.1 | 0 | 0.4 | 0 | Tr | 0 |
| 261 | **Potatoes with eggs** | 17.3 | 1.4 | 0.5 | 0.4 | 0 | 0.5 | 0 | 0 | 0.5 |
| 262 | **Quorn korma** | 15.0 | 1.3 | 0.3 | 0.3 | 0 | 0.6 | 0 | 0.2 | 0.4 |
| 263 | **Ratatouille** | 0.2 | 3.2 | 1.5 | 1.4 | 0 | 0.4 | 0 | 0 | 0.4 |
| 264 | retail | 0.1 | 3.6 | 1.7 | 1.8 | 0 | Tr | 0 | 0 | Tr |

Fibre fractions, phytic acid and fatty acids, g per 100g food
Cholesterol, mg per 100g food

| No. 15- | Food | Dietary fibre, g | | Fibre fractions, g | | | | Phytic acid g | Fatty acids, g | | | Cholesterol mg |
|---|---|---|---|---|---|---|---|---|---|---|---|---|
| | | Southgate method | Englyst method | Cellulose | Non-cellulosic polysaccharide Soluble | Non-cellulosic polysaccharide Insoluble | Lignin | | Satd | Mono-unsatd | Poly-unsatd | |
| 245 | **Pilaf**, rice with spinach | (3.5) | 1.7 | 0.7 | 0.7 | 0.4 | 0.2 | 0.08 | 0.4 | 0.6 | 1.3 | 0 |
| 246 | rice with tomato | 1.2 | 0.4 | 0.1 | 0.1 | 0.1 | 0.2 | 0.14 | 0.6 | 0.8 | 1.7 | 0 |
| 247 | **Pilau**, egg and potato | 1.3 | 0.8 | 0.2 | 0.3 | 0.2 | 0.1 | 0.08 | 0.8 | 1.2 | 1.3 | 60 |
| 248 | egg and potato, brown rice | 1.5 | 1.1 | 0.3 | 0.3 | 0.4 | Tr | N | 0.8 | 1.2 | 1.2 | 60 |
| 249 | mushroom | 1.2 | 0.4 | 0.1 | 0.1 | 0.2 | 0.2 | 0.13 | 2.5 | 1.1 | 0.5 | 9 |
| 250 | plain | 0.9 | 0.3 | 0.1 | 0.1 | 0.1 | 0.2 | 0.12 | 2.5 | 1.1 | 0.5 | 9 |
| 251 | vegetable | N | N | N | N | N | N | 0.12 | 2.4 | 1.1 | 0.5 | 9 |
| 252 | **Pizza**, cheese and tomato | 1.9 | 1.4 | 0.2 | 0.6 | 0.6 | 0.1 | 0.05 | 5.4 | 3.6 | 2.1 | 22 |
| 253 | cheese and tomato, wholemeal | 3.4 | 3.2 | 0.6 | 0.7 | 1.8 | 0.2 | 0.16 | 5.4 | 3.6 | 2.2 | 22 |
| 254 | cheese and tomato, retail, frozen | 1.9 | (1.4) | (0.2) | (0.6) | (0.6) | (0.1) | (0.05) | 4.3 | 3.6 | 1.9 | 26 |
| 255 | tomato | 2.0 | 1.4 | 0.3 | 0.6 | 0.5 | 0.2 | 0.05 | 1.5 | 7.1 | 1.4 | 0 |
| 256 | tomato, wholemeal | 3.2 | 2.9 | 0.6 | 0.7 | 1.5 | 0.3 | 0.14 | 1.5 | 7.1 | 1.5 | 0 |
| 257 | **Potato cakes**, *fried in lard* | 1.8 | 1.6 | 0.3 | 0.8 | 0.4 | Tr | 0.03 | 3.9 | 3.2 | 0.8 | 11 |
| 258 | *fried in vegetable oil* | 1.8 | 1.6 | 0.3 | 0.8 | 0.4 | Tr | 0.03 | 2.1 | 2.7 | 3.1 | 6 |
| 259 | **Potato, leek and celery bake** | 1.1 | 0.8 | 0.3 | 0.4 | 0.1 | Tr | Tr | 3.9 | 1.6 | 0.2 | 17 |
| 260 | **Potatoes**, duchesse | 1.4 | 1.2 | 0.4 | 0.7 | 0.1 | Tr | Tr | 3.0 | 1.5 | 0.3 | 49 |
| 261 | **Potatoes with eggs** | (2.0) | 1.4 | 0.5 | 0.8 | 0.1 | Tr | 0.01 | 4.4 | 6.8 | 11.0 | 172 |
| 262 | **Quorn korma** | N | N | N | N | N | N | 0.09 | 3.7 | 1.6 | 1.3 | 2 |
| 263 | **Ratatouille** | (2.0) | 1.8 | 0.6 | 0.8 | 0.3 | 0.2 | 0.02 | 0.8 | 2.4 | 3.4 | 0 |
| 264 | retail | N | (1.0) | (0.4) | N | N | N | N | 0.8 | 1.6 | 3.6 | 0 |

| No. 15- | Food | Na | K | Ca | Mg | P | Fe | Cu | Zn | Cl | Mn | Se | I |
|---|---|---|---|---|---|---|---|---|---|---|---|---|---|
| | | mg | | | | | | | | | | µg | |
| 245 | **Pilaf**, rice with spinach | 270 | 410 | 140 | 47 | 63 | 1.7 | 0.10 | 0.9 | 320 | 0.7 | (3) | (5) |
| 246 | rice with tomato | 190 | 110 | 20 | 13 | 53 | 0.3 | 0.13 | 0.6 | 310 | 0.4 | 3 | (5) |
| 247 | **Pilau**, egg and potato | 530 | 160 | 29 | 16 | 73 | 1.0 | 0.10 | 0.7 | 820 | 0.2 | (4) | 13 |
| 248 | egg and potato, brown rice | 530 | 180 | 22 | 33 | 110 | 1.1 | 0.21 | 0.7 | 870 | 0.5 | (2) | N |
| 249 | mushroom | 260 | 100 | 19 | 12 | 55 | 0.4 | 0.20 | 0.6 | 410 | 0.4 | 4 | (5) |
| 250 | plain | 270 | 73 | 21 | 13 | 48 | 0.6 | 0.12 | 0.6 | 420 | 0.4 | 3 | (5) |
| 251 | vegetable | 270 | 85 | 25 | 14 | 52 | 0.6 | 0.11 | 0.6 | 410 | 0.4 | N | (4) |
| 252 | **Pizza**, cheese and tomato | 580 | 160 | 210 | 19 | 160 | 1.1 | 0.11 | 0.8 | 950 | 0.2 | 4 | 13 |
| 253 | cheese and tomato, wholemeal | 580 | 220 | 180 | 49 | 230 | 1.6 | 0.21 | 1.5 | 930 | 1.0 | 19 | 10 |
| 254 | cheese and tomato, retail, frozen | 540 | 170 | 180 | 16 | 130 | 1.0 | 0.11 | 1.0 | 910 | 0.3 | (4) | (14) |
| 255 | tomato | 340 | 230 | 41 | 14 | 57 | 1.0 | 0.09 | 0.3 | 570 | 0.2 | 1 | 4 |
| 256 | tomato, wholemeal | 340 | 280 | 15 | 39 | 110 | 1.5 | 0.16 | 0.9 | 560 | 0.8 | 13 | 2 |
| 257 | **Potato cakes**, *fried in lard* | 170 | 240 | 49 | 16 | 57 | 0.8 | 0.09 | 0.4 | 310 | 0.2 | 2 | 7 |
| 258 | *fried in vegetable oil* | 170 | 240 | 49 | 16 | 57 | 0.8 | 0.09 | 0.4 | 310 | 0.2 | 2 | 8 |
| 259 | **Potato, leek and celery bake** | 490 | 250 | 100 | 19 | 85 | 0.6 | 0.06 | 0.5 | 790 | (0.2) | (2) | (9) |
| 260 | **Potatoes**, duchesse | 120 | 290 | 11 | 15 | 52 | 0.6 | 0.08 | 0.4 | 230 | 0.1 | 2 | 10 |
| 261 | **Potatoes with eggs** | 540 | 400 | 40 | 24 | 170 | 1.4 | 0.15 | 0.9 | 790 | 0.2 | (6) | (28) |
| 262 | **Quorn korma** | 180 | (99) | (17) | (14) | (45) | (0.4) | (0.11) | (0.4) | (230) | (0.3) | (2) | N |
| 263 | **Ratatouille** | 200 | 280 | 17 | 15 | 31 | 0.5 | 0.02 | 0.2 | 340 | 0.1 | (1) | N |
| 264 | retail | 19 | 220 | 22 | 15 | 27 | 0.6 | 0.06 | 0.2 | 55 | 0.1 | N | N |

# Vegetable Dishes

| No. 15- | Food | Retinol µg | Carotene µg | Vitamin D µg | Vitamin E mg | Thiamin mg | Ribo-flavin mg | Niacin mg | Trypt 60 mg | Vitamin B6 mg | Vitamin B12 µg | Folate µg | Panto-thenate mg | Biotin µg | Vitamin C mg |
|---|---|---|---|---|---|---|---|---|---|---|---|---|---|---|---|
| 245 | Pilaf, rice with spinach | 0 | 2630 | 0 | 1.53 | 0.11 | 0.06 | 1.4 | 0.9 | 0.16 | 0 | 58 | (0.26) | (0.6) | 10 |
| 246 | rice with tomato | 0 | 100 | 0 | 0.60 | 0.12 | 0.01 | 1.2 | 0.5 | 0.11 | 0 | 5 | (0.19) | (1.0) | 2 |
| 247 | Pilau, egg and potato | 29 | 11 | 0.3 | 0.42 | 0.06 | 0.06 | 0.6 | 1.0 | 0.12 | 0.2 | 17 | (0.33) | (3.2) | 5 |
| 248 | egg and potato, brown rice | 29 | 11 | 0.3 | 0.58 | 0.12 | 0.07 | 0.8 | 1.0 | N | 0.2 | 20 | N | N | 5 |
| 249 | mushroom | 22 | 19 | 0.1 | 0.19 | 0.11 | 0.04 | 1.3 | 0.5 | 0.10 | Tr | 6 | (0.35) | (2.0) | Tr |
| 250 | plain | 23 | 21 | 0.1 | 0.18 | 0.11 | 0.01 | 1.0 | 0.5 | 0.09 | Tr | 3 | (0.14) | (0.8) | Tr |
| 251 | vegetable | 21 | 350 | 0.1 | 0.17 | 0.11 | 0.01 | 1.0 | 0.5 | 0.09 | Tr | 6 | (0.14) | (0.7) | 1 |
| 252 | Pizza, cheese and tomato | 72 | 250 | 0.1 | 1.41 | 0.11 | 0.12 | 0.9 | 2.0 | 0.10 | 0.2 | 27 | 0.26 | 3.2 | 3 |
| 253 | cheese and tomato, wholemeal | 72 | 250 | 0.1 | 1.74 | 0.15 | 0.13 | 2.0 | 2.2 | 0.18 | 0.2 | 32 | 0.37 | 5.0 | 3 |
| 254 | cheese and tomato, retail, frozen | N | N | N | N | 0.16 | 0.14 | 0.9 | 2.0 | 0.13 | Tr | 20 | N | N | N |
| 255 | tomato | 0 | 380 | 0 | 1.35 | 0.11 | 0.04 | 1.1 | 0.6 | 0.11 | 0 | 22 | 0.23 | 2.7 | 5 |
| 256 | tomato, wholemeal | 0 | 380 | 0 | 1.62 | 0.14 | 0.05 | 2.0 | 0.8 | 0.18 | 0 | 27 | 0.33 | 4.2 | 5 |
| 257 | Potato cakes, fried in lard | 22 | 11 | Tr | 0.17 | 0.16 | 0.03 | 0.8 | 0.8 | 0.20 | Tr | 10 | 0.36 | 0.6 | 3 |
| 258 | fried in vegetable oil | 22 | 11 | Tr | 1.60 | 0.16 | 0.03 | 0.8 | 0.8 | 0.20 | Tr | 10 | 0.36 | 0.6 | 3 |
| 259 | Potato, leek and celery bake | 64 | 98 | 0.1 | 0.25 | 0.10 | 0.06 | 0.3 | 1.0 | 0.19 | 0.1 | 11 | 0.23 | 0.7 | 3 |
| 260 | Potatoes, duchesse | 59 | 21 | 0.2 | 0.27 | 0.15 | 0.05 | 0.4 | 0.8 | 0.27 | 0.2 | 15 | 0.45 | 1.8 | 3 |
| 261 | Potatoes with eggs | 85 | 2 | 0.8 | 4.46 | 0.12 | 0.25 | 1.2 | 2.2 | 0.24 | 1.1 | 18 | N | N | 4 |
| 262 | Quorn korma | 12 | 22 | Tr | 0.80 | N | 0.03 | 0.7 | N | N | Tr | 4 | 0.14 | N | 1 |
| 263 | Ratatouille | 0 | 375 | 0 | 2.07 | 0.07 | 0.01 | 0.4 | 0.2 | 0.14 | 0 | 15 | 0.12 | N | 14 |
| 264 | retail | 0 | 185 | 0 | 2.66 | 0.04 | 0.05 | 0.6 | 0.2 | 0.10 | 0 | 41 | N | N | 12 |

| No. | Food | Description and main data sources |
|-----|------|-----------------------------------|
| **15-** | | |
| 265 | **Re-fried beans** | Mashed kidney bean dish. Recipe from review of recipe collection |
| 266 | **Red pea loaf** | West Indian dish. Red kidney beans. Recipe from a personal collection |
| 267 | **Rice and black-eye beans** | West Indian dish. Recipe from review of recipe collection |
| 268 | brown rice | West Indian dish. Recipe from review of recipe collection |
| 269 | **Rice and pigeon peas** | West Indian dish. Recipe from review of recipe collection |
| 270 | brown rice | West Indian dish. Recipe from review of recipe collection |
| 271 | **Rice and red kidney beans** | West Indian dish. Recipe from review of recipe collection |
| 272 | brown rice | West Indian dish. Recipe from review of recipe collection |
| 273 | **Rice and split peas** | West Indian dish. Recipe from review of recipe collection |
| 274 | brown rice | West Indian dish. Recipe from review of recipe collection |
| 275 | **Risotto,** vegetable | White rice, vegetables, red kidney beans and cashew nuts. Recipe from Leeds Polytechnic |
| 276 | vegetable, brown rice | Rice, vegetables, red kidney beans and cashew nuts. Recipe from Leeds Polytechnic |
| 277 | **Rissoles,** chick pea, *fried in sunflower oil* | Recipe from dietary survey records |
| 278 | *fried in vegetable oil* | Recipe from dietary survey records |
| 279 | lentil, *fried in sunflower oil* | Recipe from review of recipe collection |
| 280 | *fried in vegetable oil* | Recipe from review of recipe collection |
| 281 | rice, *fried in sunflower oil* | Recipe from dietary survey records. Brown rice |
| 282 | *fried in vegetable oil* | Recipe from dietary survey records. Brown rice |
| 283 | vegetable, *fried in sunflower oil* | Potato, turnip, carrot and pea. Recipe from dietary survey records |
| 284 | *fried in vegetable oil* | Potato, turnip, carrot and pea. Recipe from dietary survey records |

**Vegetable Dishes**

Composition of food per 100g

| No. 15- | Food | Water g | Total nitrogen g | Protein g | Fat g | Carbohydrate g | Energy value kcal | kJ |
|---|---|---|---|---|---|---|---|---|
| 265 | **Re-fried beans** | 45.6 | 1.59 | 9.9 | 13.2 | 20.5 | 234 | 983 |
| 266 | **Red pea loaf** | 63.9 | 1.47 | 9.2 | 3.6 | 15.5 | 126 | 535 |
| 267 | **Rice and black-eye beans** | 56.7 | 0.95 | 5.8 | 3.6 | 34.2 | 183 | 778 |
| 268 | brown rice | 57.4 | 0.91 | 5.6 | 3.3 | 32.9 | 175 | 744 |
| 269 | **Rice and pigeon peas** | 58.5 | 0.83 | 5.1 | 3.5 | 33.3 | 175 | 745 |
| 270 | brown rice | 59.3 | 0.79 | 4.9 | 3.2 | 32.0 | 168 | 713 |
| 271 | **Rice and red kidney beans** | 57.2 | 0.91 | 5.6 | 3.5 | 32.4 | 175 | 744 |
| 272 | brown rice | 58.0 | 0.87 | 5.4 | 3.3 | 31.1 | 167 | 710 |
| 273 | **Rice and split peas** | 53.6 | 0.99 | 6.0 | 4.0 | 37.4 | 200 | 848 |
| 274 | brown rice | 54.4 | 0.95 | 5.9 | 3.7 | 36.0 | 191 | 812 |
| 275 | **Risotto**, vegetable | 66.9 | 0.78 | 4.2 | 6.5 | 19.2 | 147 | 620 |
| 276 | vegetable, brown rice | 67.2 | 0.76 | 4.1 | 6.4 | 18.6 | 143 | 603 |
| 277 | **Rissoles**, chick pea, *fried in sunflower oil* | 52.6 | 1.28 | 8.0 | 16.9 | 16.0 | 243 | 1016 |
| 278 | *fried in vegetable oil* | 52.6 | 1.28 | 8.0 | 16.9 | 16.0 | 243 | 1016 |
| 279 | lentil, *fried in sunflower oil* | 52.5 | 1.45 | 8.9 | 10.5 | 22.0 | 211 | 886 |
| 280 | *fried in vegetable oil* | 52.5 | 1.45 | 8.9 | 10.5 | 22.0 | 211 | 886 |
| 281 | rice, *fried in sunflower oil* | 51.5 | 0.95 | 5.6 | 8.8 | 30.8 | 216 | 911 |
| 282 | *fried in vegetable oil* | 51.5 | 0.95 | 5.6 | 8.8 | 30.8 | 216 | 911 |
| 283 | vegetable, *fried in sunflower oil* | 76.7 | 0.25 | 1.5 | 7.3 | 12.7 | 119 | 499 |
| 284 | *fried in vegetable oil* | 76.7 | 0.25 | 1.5 | 7.3 | 12.7 | 119 | 499 |

# Vegetable Dishes

## Carbohydrate fractions, g per 100g food

| No. 15- | Food | Starch | Total sugars | Glucose | Fructose | Galactose | Sucrose | Maltose | Lactose | Oligo-saccharides |
|---------|------|--------|--------------|---------|----------|-----------|---------|---------|---------|-------------------|
| 265 | **Re-fried beans** | 16.7 | 1.9 | 0.4 | 0.2 | 0 | 1.2 | 0 | 0.1 | 1.9 |
| 266 | **Red pea loaf** | 12.9 | 0.9 | 0.1 | Tr | 0 | 0.7 | 0 | Tr | 1.7 |
| 267 | **Rice and black-eye beans** | 32.7 | 0.8 | 0.1 | 0.1 | 0 | 0.5 | 0 | Tr | 0.7 |
| 268 | brown rice | 31.0 | 1.2 | 0.3 | 0.1 | 0 | 0.8 | 0 | Tr | 0.7 |
| 269 | **Rice and pigeon peas** | 32.3 | 0.6 | 0.1 | 0.1 | 0 | 0.4 | 0 | Tr | 0.4 |
| 270 | brown rice | 30.6 | 1.0 | 0.3 | 0.1 | 0 | 0.6 | 0 | Tr | 0.4 |
| 271 | **Rice and red kidney beans** | 31.4 | 0.5 | Tr | Tr | 0 | 0.4 | 0 | Tr | 0.5 |
| 272 | brown rice | 29.7 | 0.9 | 0.2 | Tr | 0 | 0.7 | 0 | Tr | 0.5 |
| 273 | **Rice and split peas** | 36.4 | 0.7 | 0.1 | 0.1 | 0 | 0.5 | 0 | Tr | 0.4 |
| 274 | brown rice | 34.5 | 1.1 | 0.3 | 0.1 | 0 | 0.7 | 0 | Tr | 0.4 |
| 275 | **Risotto**, vegetable | 16.2 | 2.5 | 0.6 | 0.7 | 0 | 1.2 | 0 | 0 | 0.5 |
| 276 | vegetable, brown rice | 15.4 | 2.6 | 0.6 | 0.7 | 0 | 1.2 | 0 | 0 | 0.5 |
| 277 | **Rissoles**, chick pea, *fried in sunflower oil* | 15.0 | 0.4 | Tr | Tr | 0 | 0.4 | 0 | 0 | 0.6 |
| 278 | *fried in vegetable oil* | 15.0 | 0.4 | Tr | Tr | 0 | 0.4 | 0 | 0 | 0.6 |
| 279 | lentil, *fried in sunflower oil* | 18.9 | 2.0 | N | N | 0 | N | N | 0 | 1.1 |
| 280 | *fried in vegetable oil* | 18.9 | 2.0 | N | N | 0 | N | N | 0 | 1.1 |
| 281 | rice, *fried in sunflower oil* | 29.1 | 1.5 | N | N | 0 | N | N | 0 | 0.2 |
| 282 | *fried in vegetable oil* | 29.1 | 1.5 | N | N | 0 | N | N | 0 | 0.2 |
| 283 | vegetable, *fried in sunflower oil* | 11.1 | 1.5 | 0.5 | 0.4 | 0 | 0.6 | Tr | 0 | 0.1 |
| 284 | *fried in vegetable oil* | 11.1 | 1.5 | 0.5 | 0.4 | 0 | 0.6 | Tr | 0 | 0.1 |

**Vegetable Dishes**

**Fibre fractions, phytic acid and fatty acids, g per 100g food**
**Cholesterol, mg per 100g food**

| No. 15- | Food | Dietary fibre, g Southgate method | Dietary fibre, g Englyst method | Fibre fractions, g Cellulose | Non-cellulosic polysaccharide Soluble | Non-cellulosic polysaccharide Insoluble | Lignin | Phytic acid g | Fatty acids, g Satd | Mono- unsatd | Poly- unsatd | Cholesterol mg |
|---|---|---|---|---|---|---|---|---|---|---|---|---|
| 265 | **Re-fried beans** | (10.4) | 7.0 | 1.7 | 3.1 | 2.2 | N | 0.51 | 3.1 | 4.1 | 5.3 | 1 |
| 266 | **Red pea loaf** | (8.0) | 6.0 | 1.5 | 2.9 | 1.6 | N | 0.24 | 0.9 | 1.2 | 1.1 | 40 |
| 267 | **Rice and black-eye beans** | (2.1) | 1.4 | 0.4 | 0.5 | 0.6 | 0.2 | 0.27 | 1.5 | 0.9 | 0.9 | 2 |
| 268 | brown rice | (2.4) | 1.8 | 0.5 | 0.5 | 0.9 | N | N | 1.5 | 0.8 | 0.8 | 2 |
| 269 | **Rice and pigeon peas** | 3.2 | (2.1) | N | N | N | N | N | 1.4 | 0.8 | 0.9 | 2 |
| 270 | brown rice | 3.5 | (2.6) | N | N | N | N | N | 1.4 | 0.8 | 0.8 | 2 |
| 271 | **Rice and red kidney beans** | (4.3) | 2.5 | 0.6 | 1.0 | 0.8 | N | 0.30 | 1.5 | 0.9 | 0.9 | 2 |
| 272 | brown rice | (4.6) | 2.9 | 0.7 | 1.0 | 1.1 | N | N | 1.4 | 0.8 | 0.8 | 2 |
| 273 | **Rice and split peas** | 2.7 | 1.2 | 0.2 | 0.4 | 0.6 | N | 0.14 | 1.6 | 1.0 | 1.0 | 2 |
| 274 | brown rice | 3.0 | 1.7 | 0.4 | 0.4 | 0.9 | N | N | 1.6 | 0.9 | 0.9 | 2 |
| 275 | **Risotto**, vegetable | (3.4) | 2.2 | 0.6 | 1.0 | 0.7 | N | (0.18) | 1.0 | 2.6 | 2.5 | 0 |
| 276 | vegetable, brown rice | (3.6) | 2.4 | 0.7 | 1.0 | 0.8 | N | N | 0.9 | 2.6 | 2.5 | 0 |
| 277 | **Rissoles**, chick pea, fried in sunflower oil | (4.2) | 4.1 | 1.0 | 1.4 | 1.8 | N | (0.15) | 2.1 | 3.7 | 9.7 | 25 |
| 278 | fried in vegetable oil | (4.2) | 4.1 | 1.0 | 1.4 | 1.8 | N | (0.15) | 1.9 | 5.7 | 7.8 | 25 |
| 279 | lentil, fried in sunflower oil | (4.1) | (3.6) | 1.0 | 1.0 | 1.6 | N | 0.24 | 1.4 | 2.3 | 6.0 | 21 |
| 280 | fried in vegetable oil | (4.1) | (3.6) | 1.0 | 1.0 | 1.6 | N | 0.24 | 1.3 | 3.6 | 4.7 | 21 |
| 281 | rice, fried in sunflower oil | 3.8 | (2.5) | 0.5 | 0.6 | 1.4 | 0.2 | N | 1.3 | 2.0 | 4.7 | 25 |
| 282 | fried in vegetable oil | 3.8 | (2.5) | 0.5 | 0.6 | 1.4 | 0.2 | N | 1.2 | 3.0 | 3.7 | 25 |
| 283 | vegetable, fried in sunflower oil | (1.7) | 1.5 | 0.6 | 0.8 | 0.2 | Tr | 0.01 | 1.1 | 1.6 | 4.2 | 0 |
| 284 | fried in vegetable oil | (1.7) | 1.5 | 0.6 | 0.8 | 0.2 | Tr | 0.01 | 1.0 | 2.5 | 3.4 | 0 |

| No. Food | Na | K | Ca | Mg | P | Fe | Cu | Zn | Cl | Mn | Se | I |
|---|---|---|---|---|---|---|---|---|---|---|---|---|
| 15- | | | | | mg | | | | | | µg | |
| 265 **Re-fried beans** | 390 | 630 | 48 | 69 | 190 | 3.0 | 0.31 | 1.4 | 580 | 0.6 | 7 | (5) |
| 266 **Red pea loaf** | 220 | 390 | 51 | 43 | 140 | 2.5 | 0.22 | 1.1 | 330 | 0.5 | 7 | N |
| 267 **Rice and black-eye beans** | 150 | 240 | 28 | 32 | 110 | 1.4 | 0.23 | 1.0 | 220 | 0.6 | 4 | N |
| 268 brown rice | 140 | 270 | 16 | 55 | 160 | 1.6 | 0.38 | 1.0 | 290 | 0.9 | 2 | N |
| 269 **Rice and pigeon peas** | 140 | 260 | 35 | 25 | 87 | 0.7 | 0.29 | 0.9 | 210 | 0.5 | N | N |
| 270 brown rice | 140 | 280 | 24 | 47 | 130 | 1.0 | 0.43 | 0.9 | 280 | 0.8 | N | N |
| 271 **Rice and red kidney beans** | 150 | 260 | 30 | 33 | 110 | 1.2 | 0.22 | 1.0 | 220 | 0.6 | 5 | N |
| 272 brown rice | 150 | 290 | 18 | 57 | 160 | 1.4 | 0.36 | 1.0 | 290 | 0.9 | 3 | N |
| 273 **Rice and split peas** | 160 | 210 | 23 | 33 | 95 | 1.1 | (0.16) | (1.2) | 250 | 0.6 | N | N |
| 274 brown rice | 160 | 240 | 10 | 58 | 150 | 1.4 | (0.31) | (1.2) | 320 | 1.0 | N | N |
| 275 **Risotto**, vegetable | 330 | 240 | 31 | 29 | 100 | 1.1 | 0.30 | 0.8 | 420 | 0.4 | 6 | N |
| 276 vegetable, brown rice | 330 | 260 | 25 | 40 | 130 | 1.2 | 0.37 | 0.8 | 460 | 0.5 | 5 | N |
| 277 **Rissoles**, chick pea, *fried in* | | | | | | | | | | | | |
| *sunflower oil* | 230 | 130 | 48 | 24 | 94 | 1.7 | 0.06 | 0.9 | 290 | 0.8 | 2 | N |
| 278 *fried in vegetable oil* | 230 | 130 | 48 | 24 | 94 | 1.7 | 0.06 | 0.9 | 290 | 0.8 | 2 | N |
| 279 lentil, *fried in sunflower oil* | 230 | 330 | 39 | 45 | 140 | 3.6 | 0.33 | 1.5 | 370 | 0.7 | (34) | (4) |
| 280 *fried in vegetable oil* | 230 | 330 | 39 | 45 | 140 | 3.6 | 0.33 | 1.5 | 370 | 0.7 | (34) | (5) |
| 281 rice, *fried in sunflower oil* | 350 | 170 | 32 | 51 | 150 | 1.7 | 0.26 | 1.1 | 590 | 1.2 | (14) | N |
| 282 *fried in vegetable oil* | 350 | 170 | 32 | 51 | 150 | 1.7 | 0.26 | 1.1 | 590 | 1.2 | (14) | N |
| 283 vegetable, *fried in sunflower oil* | N | 220 | 16 | 10 | 29 | 0.4 | 0.05 | 0.2 | N | 0.1 | (1) | (3) |
| 284 *fried in vegetable oil* | N | 220 | 16 | 10 | 29 | 0.4 | 0.05 | 0.2 | N | 0.1 | (1) | (3) |

**Vegetable Dishes**

| No. 15- | Food | Retinol µg | Carotene µg | Vitamin D µg | Vitamin E mg | Thiamin mg | Ribo-flavin mg | Niacin mg | Trypt 60 mg | Vitamin B₆ mg | Vitamin B₁₂ µg | Folate µg | Panto-thenate mg | Biotin µg | Vitamin C mg |
|---|---|---|---|---|---|---|---|---|---|---|---|---|---|---|---|
| 265 | **Re-fried beans** | 110 | 195 | 1.1 | 0.60 | 0.24 | 0.07 | 0.8 | 1.6 | 0.16 | 0 | 29 | 0.28 | N | 1 |
| 266 | **Red pea loaf** | 24 | 7 | 0.2 | 0.66 | 0.13 | 0.08 | 0.4 | 1.6 | 0.10 | 0.3 | 21 | 0.30 | N | Tr |
| 267 | **Rice and black-eye beans** | 12 | 17 | 0.1 | 0.19 | 0.21 | 0.03 | 1.3 | 1.3 | 0.13 | 0 | 50 | (0.33) | (2.9) | Tr |
| 268 | brown rice | 12 | 17 | 0.1 | 0.40 | 0.25 | 0.04 | 1.6 | 1.2 | N | 0 | 54 | N | N | Tr |
| 269 | **Rice and pigeon peas** | 12 | 16 | 0.1 | (0.19) | 0.20 | 0.02 | 1.3 | 0.7 | N | 0 | 10 | N | N | Tr |
| 270 | brown rice | 12 | 16 | 0.1 | (0.39) | 0.24 | 0.03 | 1.5 | 0.7 | N | 0 | 14 | N | N | Tr |
| 271 | **Rice and red kidney beans** | 12 | 14 | 0.1 | 0.26 | 0.18 | 0.03 | 1.3 | 1.0 | 0.12 | 0 | 12 | (0.24) | N | Tr |
| 272 | brown rice | 12 | 14 | 0.1 | 0.47 | 0.22 | 0.04 | 1.5 | 1.0 | N | 0 | 17 | N | N | Tr |
| 273 | **Rice and split peas** | 13 | 36 | 0.1 | 0.37 | 0.20 | 0.03 | 1.5 | 1.1 | 0.11 | 0 | N | N | 0.8 | Tr |
| 274 | brown rice | 13 | 36 | 0.1 | N | 0.25 | 0.04 | 1.8 | 1.1 | N | 0 | N | N | N | Tr |
| 275 | **Risotto,** vegetable | 0 | 515 | 0 | 1.14 | 0.14 | 0.07 | 1.4 | 0.8 | 0.16 | 0 | 10 | 0.50 | N | 10 |
| 276 | vegetable, brown rice | 0 | 515 | 0 | (1.25) | 0.16 | 0.08 | 1.5 | 0.8 | (0.12) | 0 | 12 | (0.43) | N | 10 |
| 277 | **Rissoles,** chick pea, *fried in sunflower oil* | 12 | 65 | 0.1 | 8.16 | 0.05 | 0.06 | 0.2 | 1.3 | 0.04 | 0.2 | 7 | N | N | 2 |
| 278 | *fried in vegetable oil* | 12 | 65 | 0.1 | 4.83 | 0.05 | 0.06 | 0.2 | 1.3 | 0.04 | 0.2 | 7 | N | N | 2 |
| 279 | *lentil, fried in sunflower oil* | 10 | 835 | 0.1 | (4.51) | 0.15 | 0.09 | 1.1 | 1.5 | 0.24 | 0.1 | 20 | (0.20) | (1.9) | 1 |
| 280 | *fried in vegetable oil* | 10 | 835 | 0.1 | (2.30) | 0.15 | 0.09 | 1.1 | 1.5 | 0.24 | 0.1 | 20 | (0.20) | (1.9) | 1 |
| 281 | *rice, fried in sunflower oil* | 12 | 1 | 0.1 | 3.51 | 0.15 | 0.07 | 2.2 | 1.3 | N | 0.2 | 10 | N | N | Tr |
| 282 | *fried in vegetable oil* | 12 | 1 | 0.1 | 1.87 | 0.15 | 0.07 | 2.2 | 1.3 | N | 0.2 | 10 | N | N | Tr |
| 283 | *vegetable, fried in sunflower oil* | 15 | 1160 | 0.2 | 2.90 | 0.11 | 0.01 | 0.4 | 0.3 | 0.16 | 0 | 9 | 0.28 | 0.3 | 4 |
| 284 | *fried in vegetable oil* | 15 | 1160 | 0.2 | 1.48 | 0.11 | 0.01 | 0.4 | 0.3 | 0.16 | 0 | 9 | 0.28 | 0.3 | 4 |

No. Food
15-

Description and main data sources

| No. 15- | Food | Description and main data sources |
|---|---|---|
| 285 | **Roulade**, spinach | Recipe from dietary survey records |
| 286 | **Salad**, bean, retail | Assorted beans in French dressing. Recipe from manufacturer |
| 287 | beetroot | Beetroot and onion in French dressing. Recipe from review of recipe collection |
| 288 | carrot and nut with French dressing, retail | Recipe from manufacturer |
| 289 | carrot and nut with mayonnaise, retail | Recipe from manufacturer |
| 290 | Florida, retail | White cabbage, celery and fruit. Recipe from manufacturer |
| 291 | Greek | Salad with feta cheese, olives and olive oil. Recipe from review of recipe collection |
| 292 | green | Lettuce, cucumber, pepper and celery. Recipe from review of recipe collection |
| 293 | pasta | Pasta, vegetables and mayonnaise. Recipe from dietary survey records |
| 294 | pasta, wholemeal | Pasta, vegetables and mayonnaise. Recipe from dietary survey records |
| 295 | potato, with French dressing | Recipe from review of recipe collection |
| 296 | potato, with mayonnaise | Recipe from review of recipe collection |
| 297 | potato, with mayonnaise, retail | Recipe from manufacturers |
| 298 | potato, with reduced calorie dressing, retail | Recipe from manufacturer |
| 299 | rice | Rice, vegetables, nut and raisin. Recipe from dietary survey records |
| 300 | rice, brown | Rice, vegetables, nut and raisin. Recipe from dietary survey records |
| 301 | tomato and onion | Recipe from review of recipe collection |
| 302 | vegetable, canned | Recipe from Wiles et al, 1980. Dissection of shop-bought samples |

**Vegetable Dishes**

15-285 to 15-302

**Composition of food per 100g**

| No. 15- | Food | Water g | Total nitrogen g | Protein g | Fat g | Carbohydrate g | Energy value kcal | kJ |
|---|---|---|---|---|---|---|---|---|
| 285 | **Roulade**, spinach | 68.0 | 1.60 | 10.1 | 13.7 | 4.6 | 180 | 749 |
| 286 | **Salad**, bean, retail | 69.9 | 0.68 | 4.2 | 9.3 | 12.8 | 147 | 618 |
| 287 | beetroot | 77.8 | 0.32 | 2.0 | 6.8 | 8.4 | 100 | 417 |
| 288 | carrot and nut with French dressing, retail | 62.6 | 0.38 | 2.1 | 17.6 | 13.7 | 218 | 903 |
| 289 | carrot and nut with mayonnaise, retail | 57.5 | 0.71 | 4.0 | 24.8 | 10.4 | 278 | 1056 |
| 290 | Florida, retail | 66.0 | 0.15 | 0.9 | 20.5 | 9.7 | 224 | 832 |
| 291 | Greek | 79.4 | 0.43 | 2.7 | 12.5 | 1.9 | 130 | 540 |
| 292 | green | 95.1 | 0.12 | 0.7 | 0.3 | 1.8 | 12 | 51 |
| 293 | pasta | 75.4 | 0.44 | 2.6 | 7.4 | 13.3 | 127 | 498 |
| 294 | pasta, wholemeal | 73.1 | 0.52 | 3.1 | 7.5 | 13.8 | 131 | 519 |
| 295 | potato, with French dressing | 72.0 | 0.25 | 1.5 | 11.2 | 13.7 | 157 | 656 |
| 296 | potato, with mayonnaise | 64.0 | 0.26 | 1.6 | 20.8 | 12.2 | 239 | 891 |
| 297 | potato, with mayonnaise, retail | 59.0 | 0.25 | 1.5 | 26.5 | 11.4 | 287 | 1063 |
| 298 | potato, with reduced calorie dressing, retail | 78.9 | 0.21 | 1.3 | 4.1 | 14.8 | 97 | 410 |
| 299 | rice | 65.8 | 0.54 | 3.1 | 7.5 | 23.1 | 166 | 698 |
| 300 | rice, brown | 64.7 | 0.53 | 3.1 | 7.4 | 23.7 | 167 | 703 |
| 301 | tomato and onion | 86.4 | 0.13 | 0.8 | 6.1 | 4.0 | 72 | 302 |
| 302 | vegetable, canned | 72.9 | 0.26 | 1.6 | 9.8 | 13.0 | 143 | 598 |

# Vegetable Dishes

**Carbohydrate fractions, g per 100g food**

| No. 15- | Food | Starch | Total sugars | Individual sugars | | | | | | Oligo-saccharides |
|---|---|---|---|---|---|---|---|---|---|---|
| | | | | Glucose | Fructose | Galactose | Sucrose | Maltose | Lactose | |
| 285 | **Roulade**, spinach | 2.5 | 2.1 | 0.2 | 0.2 | 0 | 0.1 | Tr | 1.6 | 0 |
| 286 | **Salad**, bean, retail | 9.7 | 2.6 | 0.4 | 0.4 | 0 | 1.8 | 0 | 0 | 0.5 |
| 287 | beetroot | 0.5 | 7.6 | 0.4 | 0.3 | 0 | 6.8 | 0 | 0 | 0.3 |
| 288 | carrot and nut with French dressing, retail | 0.4 | 13.2 | 5.0 | 4.7 | 0 | 3.5 | 0 | 0 | 0.1 |
| 289 | carrot and nut with mayonnaise, retail | 0.9 | 9.5 | 3.1 | 3.0 | 0.4 | 2.3 | 0 | 0.7 | 0.1 |
| 290 | Florida, retail | 0.1 | 9.6 | 3.2 | 2.9 | 0 | 2.3 | 1.2 | 0 | Tr |
| 291 | Greek | Tr | 1.9 | 0.8 | 0.9 | 0 | 0 | 0 | 0.2 | Tr |
| 292 | green | 0.1 | 1.7 | 0.8 | 0.9 | 0 | 0.1 | 0 | 0 | Tr |
| 293 | pasta | 11.0 | 2.2 | 0.6 | 0.6 | 0 | 0.8 | 0.1 | 0 | 0.2 |
| 294 | pasta, wholemeal | 11.1 | 2.6 | 0.7 | 0.6 | 0 | 0.9 | 0.2 | 0 | 0.2 |
| 295 | potato, with French dressing | 12.6 | 0.9 | 0.3 | 0.2 | 0 | 0.4 | 0 | 0 | 0.1 |
| 296 | potato, with mayonnaise | 10.9 | 1.1 | 0.3 | 0.2 | 0 | 0.7 | 0 | 0 | 0.1 |
| 297 | potato, with mayonnaise, retail | 10.2 | 1.1 | 0.2 | 0.1 | 0 | 0.7 | 0 | 0 | 0.1 |
| 298 | potato, with reduced calorie dressing, retail | 10.8 | 4.0 | 0.2 | 0.1 | 0 | 3.8 | 0 | 0 | Tr |
| 299 | rice | 18.2 | 4.8 | 1.9 | 1.9 | 0 | 1.0 | Tr | 0 | Tr |
| 300 | rice, brown | 18.6 | 5.1 | 2.0 | 1.9 | 0 | 1.2 | Tr | 0 | Tr |
| 301 | tomato and onion | Tr | 3.5 | 1.5 | 1.4 | 0 | 0.5 | 0 | 0 | 0.6 |
| 302 | vegetable, canned | 5.9 | 6.9 | 0.8 | 0.8 | 0 | 5.2 | Tr | 0 | 0.3 |

**Fibre fractions, phytic acid and fatty acids, g per 100g food**
**Cholesterol, mg per 100g food**

| No. 15- | Food | Dietary fibre, g Southgate method | Englyst method | Cellulose | Non-cellulosic polysaccharide Soluble | Insoluble | Lignin | Phytic acid g | Satd | Fatty acids, g Mono-unsatd | Poly-unsatd | Cholesterol mg |
|---|---|---|---|---|---|---|---|---|---|---|---|---|
| 285 | **Roulade**, spinach | 1.6 | 1.1 | 0.4 | 0.4 | 0.2 | Tr | Tr | 7.5 | 4.0 | 0.9 | 133 |
| 286 | **Salad**, bean, retail | (3.7) | 3.0 | 0.9 | N | N | N | (0.12) | 1.0 | 3.4 | 4.2 | 0 |
| 287 | beetroot | 2.0 | 1.7 | 0.5 | 0.8 | 0.3 | 0.1 | Tr | 0.7 | 2.5 | 3.2 | 0 |
| 288 | carrot and nut with French dressing, retail | 3.2 | 2.4 | 0.8 | 1.2 | 0.3 | N | 0.11 | 1.7 | 9.2 | 5.8 | 0 |
| 289 | carrot and nut with mayonnaise, retail | 2.2 | 1.8 | 0.6 | 0.9 | 0.3 | N | 0.08 | 3.9 | 6.7 | 13.1 | 20 |
| 290 | Florida, retail | 1.2 | 1.0 | 0.3 | 0.4 | 0.2 | 0.1 | Tr | 3.0 | 4.7 | 11.9 | 20 |
| 291 | Greek | (1.1) | 0.8 | 0.3 | 0.3 | 0.2 | 0.1 | 0.02 | (3.3) | (7.2) | (1.3) | 9 |
| 292 | green | (1.2) | 1.0 | 0.4 | 0.4 | 0.2 | 0.1 | (0.03) | Tr | Tr | 0.1 | 0 |
| 293 | pasta | (2.1) | 1.6 | 0.4 | 0.7 | 0.4 | 0.2 | N | 1.1 | 1.6 | 4.3 | 6 |
| 294 | pasta, wholemeal | (3.2) | 2.7 | 0.7 | 0.8 | 1.1 | 0.2 | N | 1.1 | 1.6 | 4.3 | 6 |
| 295 | potato, with French dressing | 1.3 | 1.1 | 0.4 | 0.6 | 0.1 | Tr | Tr | 1.2 | 4.2 | 5.1 | 0 |
| 296 | potato, with mayonnaise | 1.1 | 0.9 | 0.3 | 0.5 | 0.1 | Tr | Tr | 3.0 | 4.7 | 12.0 | 20 |
| 297 | potato, with mayonnaise, retail | 0.9 | 0.8 | 0.3 | 0.5 | 0.1 | Tr | Tr | 3.9 | 6.1 | 15.4 | 26 |
| 298 | potato, with reduced calorie dressing, retail | 0.9 | 0.8 | 0.3 | 0.5 | 0.1 | Tr | Tr | 0.4 | 1.4 | 1.9 | 1 |
| 299 | rice | (1.6) | 0.7 | 0.2 | 0.2 | 0.3 | N | 0.13 | 1.1 | 3.0 | 2.9 | 0 |
| 300 | rice, brown | (1.9) | 1.1 | 0.3 | 0.2 | 0.5 | N | N | 1.1 | 3.0 | 2.8 | 0 |
| 301 | tomato and onion | 1.2 | 1.0 | 0.4 | 0.5 | 0.2 | 0.2 | 0.02 | 0.7 | 2.3 | 2.9 | 0 |
| 302 | vegetable, canned | (1.8) | (1.2) | 0.6 | 0.4 | 0.1 | Tr | 0.01 | 1.3 | 2.0 | 6.1 | 13 |

# Vegetable Dishes

## Inorganic constituents per 100g food

| No. 15- | Food | Na | K | Ca | Mg | P | Fe (mg) | Cu | Zn | Cl | Mn | Se (µg) | I |
|---|---|---|---|---|---|---|---|---|---|---|---|---|---|
| 285 | **Roulade**, spinach | 390 | 210 | 260 | 28 | 180 | 1.4 | 0.04 | 1.1 | 530 | 0.3 | (6) | (29) |
| 286 | **Salad**, bean, retail | 280 | 190 | 36 | 24 | 74 | 1.5 | 0.07 | 0.5 | 410 | 0.3 | 2 | N |
| 287 | beetroot | 150 | 420 | 26 | 14 | 72 | 0.7 | 0.03 | 0.4 | N | 0.7 | Tr | N |
| 288 | carrot and nut with French dressing, retail | 130 | 300 | 38 | 23 | 50 | 0.8 | 0.18 | 0.3 | 200 | 0.6 | 1 | 3 |
| 289 | carrot and nut with mayonnaise, retail | 140 | 250 | 49 | 27 | 84 | 0.6 | 0.14 | 0.6 | 230 | 0.3 | N | 21 |
| 290 | Florida, retail | 130 | 120 | 25 | 5 | 20 | 0.4 | 0.02 | 0.1 | 220 | 0.2 | N | N |
| 291 | Greek | 360 | 150 | 61 | 9 | 61 | 0.4 | 0.03 | (0.2) | 600 | (0.1) | N | N |
| 292 | green | 6 | 160 | 19 | 7 | 33 | 0.4 | 0.01 | 0.1 | 31 | 0.2 | Tr | 2 |
| 293 | pasta | (51) | 84 | 14 | 11 | 37 | 0.6 | (0.06) | 0.4 | (86) | 0.2 | Tr | (4) |
| 294 | pasta, wholemeal | (73) | 140 | 16 | 24 | 69 | 1.1 | (0.10) | 0.7 | (120) | 0.5 | N | N |
| 295 | potato, with French dressing | 120 | 240 | 9 | 13 | 28 | 0.5 | 0.06 | 0.3 | 220 | 0.1 | 1 | 3 |
| 296 | potato, with mayonnaise | 130 | 210 | 8 | 10 | 30 | 0.4 | 0.05 | 0.2 | 240 | 0.1 | N | 12 |
| 297 | potato, with mayonnaise, retail | 160 | 180 | 6 | 9 | 29 | 0.4 | 0.05 | 0.2 | 290 | 0.1 | N | 14 |
| 298 | potato, with reduced calorie dressing, retail | 140 | 190 | 4 | 11 | 23 | 0.3 | 0.05 | 0.2 | 230 | 0.1 | 1 | 3 |
| 299 | rice | 190 | 180 | 21 | 24 | 74 | 1.0 | 0.20 | 0.8 | 35 | 0.2 | (4) | (4) |
| 300 | rice, brown | 190 | 210 | 13 | 42 | 110 | 1.2 | 0.30 | 0.8 | 82 | 0.6 | (2) | (1) |
| 301 | tomato and onion | 67 | 210 | 11 | 6 | 24 | 0.4 | 0.02 | 0.1 | 140 | 0.1 | Tr | 2 |
| 302 | vegetable, canned | 530 | 130 | 23 | 10 | 36 | 0.8 | 0.03 | 0.2 | 810 | 0.1 | N | N |

| No. 15- | Food | Retinol µg | Carotene µg | Vitamin D µg | Vitamin E mg | Thiamin mg | Ribo-flavin mg | Niacin mg | Trypt 60 mg | Vitamin B6 mg | Vitamin B12 µg | Folate µg | Panto-thenate mg | Biotin µg | Vitamin C mg |
|---|---|---|---|---|---|---|---|---|---|---|---|---|---|---|---|
| 285 | **Roulade**, spinach | 150 | 1855 | 0.6 | (1.30) | 0.06 | 0.22 | 0.4 | 2.6 | 0.09 | 1.0 | (31) | 0.60 | 5.3 | 2 |
| 286 | **Salad**, bean, retail | 0 | 250 | 0 | 0.27 | 0.09 | 0.04 | 0.5 | 0.8 | 0.09 | 0 | 44 | (0.12) | N | 6 |
| 287 | beetroot | 0 | 22 | 0 | 0.04 | 0.03 | 0.01 | 0.2 | 0.3 | 0.06 | 0 | 86 | 0.09 | 0.1 | 5 |
| 288 | carrot and nut with French dressing, retail | 0 | 5165 | 0 | 2.95 | 0.12 | 0.03 | 0.3 | 0.5 | 0.17 | 0 | 17 | 0.32 | 8.5 | 4 |
| 289 | carrot and nut with mayonnaise, retail | 23 | 3550 | 0.1 | 6.28 | 0.18 | 0.07 | 1.6 | 0.9 | 0.15 | 0.2 | 21 | 0.45 | N | 3 |
| 290 | Florida, retail | 23 | 57 | 0.1 | 5.22 | 0.06 | 0.03 | 0.2 | 0.2 | 0.08 | 0.1 | 13 | 0.11 | N | 14 |
| 291 | Greek | 30 | 260 | 0.1 | 1.12 | 0.04 | 0.04 | 0.4 | 0.6 | 0.11 | 0.2 | 16 | 0.21 | N | 26 |
| 292 | green | 0 | 195 | 0 | 0.41 | 0.05 | 0.01 | 0.2 | 0.1 | 0.11 | 0 | 29 | (0.21) | N | 36 |
| 293 | pasta | 7 | 1365 | Tr | 1.95 | 0.04 | 0.02 | 0.4 | 0.5 | 0.09 | Tr | 16 | N | N | 21 |
| 294 | pasta, wholemeal | 7 | 1365 | Tr | 1.95 | 0.14 | 0.03 | 0.8 | 0.6 | 0.12 | Tr | 17 | N | N | 21 |
| 295 | potato, with French dressing | 0 | 50 | 0 | 0.09 | 0.15 | 0.01 | 0.4 | 0.3 | 0.27 | 0 | 23 | 0.30 | 0.3 | 7 |
| 296 | potato, with mayonnaise | 23 | 70 | 0.1 | 5.26 | 0.13 | 0.03 | 0.4 | 0.4 | 0.23 | 0.1 | 21 | N | N | 6 |
| 297 | potato, with mayonnaise, retail | 30 | 35 | 0.1 | 6.68 | 0.12 | 0.03 | 0.3 | 0.4 | 0.21 | 0.2 | 18 | N | N | 4 |
| 298 | potato, with reduced calorie dressing, retail | Tr | Tr | Tr | 1.01 | 0.12 | 0.01 | 0.3 | 0.3 | 0.22 | 0 | 17 | 0.26 | 0.3 | 4 |
| 299 | rice | 0 | 135 | 0 | (1.19) | 0.05 | 0.02 | 0.9 | 0.7 | 0.12 | 0 | 18 | (0.14) | (1.2) | 12 |
| 300 | rice, brown | 0 | 135 | 0 | (1.35) | 0.12 | 0.03 | 1.1 | 0.7 | N | 0 | 21 | N | N | 12 |
| 301 | tomato and onion | 0 | 435 | 0 | 0.90 | 0.09 | 0.01 | 0.9 | 0.1 | 0.14 | 0 | 15 | 0.20 | 1.2 | 13 |
| 302 | vegetable, canned | 2 | 755 | 0.1 | 3.43 | (0.02) | (0.02) | (0.4) | 0.3 | N | 0.2 | 8 | N | N | 2 |

| No. | Food | Description and main data sources |
|-----|------|-----------------------------------|
| **15-** | | |
| 303 | **Salad, Waldorf** | Apple, walnut and celery with mayonnaise. Recipe from review of recipe collection |
| 304 | **Waldorf**, retail | Recipe from manufacturer |
| 305 | **Samosas**, vegetable, retail | 5 samples, 3 brands; mixed vegetable filling |
| 306 | **Sauce**, curry, onion, *with butter* | Onion base for curry. Recipe from review of recipe collection |
| 307 | *with vegetable oil* | Onion base for curry. Recipe from review of recipe collection |
| 308 | curry, sweet | Basic UK-type sauce with fruit and coconut. Recipe from review of recipe collection |
| 309 | curry, tomato and onion | Gujerati dish. Recipe from a personal collection |
| 310 | tomato base | Sauce for pasta or pizza. Recipe from review of recipe collection |
| 311 | tomato and mushroom | Sauce for pasta or pizza. Recipe from review of recipe collection |
| 312 | **Shepherd's pie**, vegetable | Vegetable and lentil base with potato topping. Recipe from review of recipe collection |
| 313 | vegetable, retail | Vegetable, lentil and barley base with potato topping. Recipe from manufacturer |
| 314 | **Soya mince**, granules | 10 samples of the same brand (Cross and Blackwell) |
| 315 | **Sweet potato and onion layer** | West Indian dish. Recipe from a personal collection |
| 316 | **Tabouleh** | Bulgur wheat based salad. Recipe from review of recipe collection |
| 317 | **Tagliatelle**, with vegetables, retail | Recipe from manufacturer |
| 318 | **Tempeh burgers**, *fried in vegetable oil* | Tempeh and rice burgers. Recipe from review of recipe collection |
| 319 | **Tofu burger**, *baked* | Tofu, oat and vegetable. Recipe from Leeds Polytechnic |
| 320 | **Tofu spread** | Tofu, vegetable and mayonnaise paste. Recipe from Leeds Polytechnic |

## Vegetable Dishes

### Composition of food per 100g

| No. 15- | Food | Water g | Total nitrogen g | Protein g | Fat g | Carbohydrate g | Energy value kcal | Energy value kJ |
|---|---|---|---|---|---|---|---|---|
| 303 | **Salad**, Waldorf | 69.8 | 0.25 | 1.4 | 17.7 | 7.5 | 193 | 735 |
| 304 | Waldorf, retail | 45.1 | 0.37 | 2.1 | 39.6 | 8.6 | 397 | 1476 |
| 305 | **Samosas**, vegetable, retail | 51.3 | 0.82 | 5.1 | 9.3 | 30.0 | 217 | 911 |
| 306 | **Sauce**, curry, onion, *with butter* | 64.6 | 0.21 | 1.3 | 24.7 | 6.4 | 251 | 1034 |
| 307 | *with vegetable oil* | 59.9 | 0.19 | 1.2 | 30.1 | 6.4 | 299 | 1232 |
| 308 | curry, sweet | 81.0 | 0.20 | 1.2 | 5.6 | 9.6 | 91 | 380 |
| 309 | curry, tomato and onion | 68.4 | 0.27 | 1.7 | 19.1 | 6.1 | 198 | 819 |
| 310 | tomato base | 85.4 | 0.23 | 1.4 | 5.0 | 5.5 | 71 | 297 |
| 311 | tomato and mushroom | 85.3 | 0.30 | 1.6 | 3.5 | 5.8 | 59 | 251 |
| 312 | **Shepherd's pie**, vegetable | 71.7 | 0.76 | 4.6 | 4.0 | 15.8 | 113 | 476 |
| 313 | vegetable, retail | 77.5 | 0.30 | 1.9 | 4.9 | 13.3 | 101 | 425 |
| 314 | **Soya mince**, granules | 4.6 | 6.91 | 43.2 | 5.4 | 11.0 | 263 | 1110 |
| 315 | **Sweet potato and onion layer** | 68.4 | 0.56 | 3.5 | 2.9 | 22.2 | 123 | 519 |
| 316 | **Tabouleh** | 73.7 | 0.45 | 2.6 | 4.6 | 17.2 | 119 | 496 |
| 317 | **Tagliatelle**, with vegetables, retail | 83.3 | 0.28 | 1.6 | 3.0 | 11.0 | 74 | 315 |
| 318 | **Tempeh burgers**, *fried in vegetable oil* | 59.5 | 1.25 | 7.7 | 8.1 | 23.8 | 194 | 813 |
| 319 | **Tofu burger**, *baked* | 71.0 | 1.34 | 8.2 | 4.3 | 13.3 | 118 | 498 |
| 320 | **Tofu spread** | 68.9 | 0.74 | 4.6 | 20.2 | 2.4 | 208 | 775 |

| No. 15- | Food | Starch | Total sugars | Glucose | Fructose | Galactose | Sucrose | Maltose | Lactose | Oligo-saccharides |
|---|---|---|---|---|---|---|---|---|---|---|
| 303 | **Salad**, Waldorf | 0.1 | 7.4 | 1.1 | 3.7 | 0 | 2.6 | 0 | 0 | 0 |
| 304 | Waldorf, retail | 0.2 | 8.4 | 3.3 | 3.9 | 0 | 1.2 | 0 | 0 | 0 |
| 305 | **Samosas**, vegetable, retail | 27.3 | 2.7 | 0.3 | 0.2 | 0 | 1.1 | 1.1 | 0 | Tr |
| 306 | **Sauce**, curry, onion, *with butter* | 0.5 | 4.2 | 1.6 | 1.2 | 0 | 1.4 | 0 | Tr | 1.7 |
| 307 | *with vegetable oil* | 0.5 | 4.2 | 1.6 | 1.2 | 0 | 1.4 | 0 | 0 | 1.7 |
| 308 | curry, sweet | 3.6 | 5.5 | 2.1 | 2.5 | 0 | 0.9 | Tr | 0 | 0.5 |
| 309 | curry, tomato and onion | 0.6 | 4.6 | 1.9 | 1.8 | 0 | 0.7 | Tr | 0 | 0.9 |
| 310 | tomato base | 0.3 | 4.8 | 1.9 | 1.9 | 0 | 0.9 | 0 | 0 | 0.4 |
| 311 | tomato and mushroom | 0.3 | 4.8 | 2.0 | 1.9 | 0 | 0.7 | 0 | 0 | 0.8 |
| 312 | **Shepherd's pie**, vegetable | 14.0 | 1.4 | 0.5 | 0.4 | 0 | 0.6 | 0 | 0 | 0.4 |
| 313 | vegetable, retail | 11.4 | 1.7 | 0.6 | 0.5 | 0 | 0.5 | Tr | 0.1 | 0.1 |
| 314 | **Soya mince**, granules | 5.6 | 5.4 | Tr | 0.1 | 0 | 5.4 | 0 | 0 | 0 |
| 315 | **Sweet potato and onion layer** | 10.9 | 11.0 | N | N | 0 | N | N | 1.8 | 0.3 |
| 316 | **Tabouleh** | 16.5 | 0.7 | 0.3 | 0.3 | 0 | 0 | 0 | 0 | 0 |
| 317 | **Tagliatelle**, with vegetables, retail | 8.6 | 2.2 | 0.5 | 0.5 | 0 | 0.9 | 0.1 | 0.3 | 0.1 |
| 318 | **Tempeh burgers**, *fried in vegetable oil* | 22.2 | 1.2 | 0.4 | 0.2 | 0 | 0.6 | Tr | 0 | 0.5 |
| 319 | **Tofu burger**, *baked* | 10.0 | 2.8 | 0.7 | 0.4 | 0 | 1.6 | Tr | 0 | 0.4 |
| 320 | **Tofu spread** | 0.2 | 1.6 | 0.5 | 0.4 | 0 | 0.6 | 0 | 0 | 0.3 |

| No. 15- | Food | Dietary fibre, g Southgate method | Dietary fibre, g Englyst method | Cellulose | Non-cellulosic polysaccharide Soluble | Non-cellulosic polysaccharide Insoluble | Lignin | Phytic acid g | Fatty acids, g Satd | Fatty acids, g Mono-unsatd | Fatty acids, g Poly-unsatd | Cholesterol mg |
|---|---|---|---|---|---|---|---|---|---|---|---|---|
| 303 | **Salad, Waldorf** | 1.7 | 1.3 | 0.5 | 0.5 | 0.3 | Tr | (0.04) | 2.3 | 3.8 | 10.7 | 13 |
| 304 | **Waldorf**, retail | 1.6 | 1.0 | 0.4 | 0.4 | 0.2 | N | (0.05) | 5.5 | 8.8 | 23.6 | 33 |
| 305 | **Samosas**, vegetable, retail | N | 2.5 | 0.6 | 1.2 | 0.7 | N | N | N | N | N | 0 |
| 306 | **Sauce**, curry, onion, *with butter* | 1.6 | 1.5 | 0.5 | 0.7 | 0.4 | N | 0.02 | 16.1 | 5.9 | 0.9 | 68 |
| 307 | *with vegetable oil* | 1.6 | 1.5 | 0.5 | 0.7 | 0.4 | N | 0.02 | 3.1 | 10.6 | 14.4 | 0 |
| 308 | curry, sweet | 1.8 | 1.4 | 0.4 | 0.4 | 0.6 | 0.1 | 0.01 | 1.5 | 1.5 | 2.0 | 0 |
| 309 | curry, tomato and onion | 1.3 | 1.1 | 0.4 | 0.6 | 0.2 | Tr | 0.02 | 2.0 | 6.7 | 9.1 | 0 |
| 310 | tomato base | 1.2 | 1.0 | 0.4 | 0.4 | 0.1 | Tr | 0.01 | 0.7 | 3.4 | 0.6 | 0 |
| 311 | tomato and mushroom | 1.6 | 1.4 | 0.5 | 0.6 | 0.2 | Tr | 0.01 | 0.4 | 0.8 | 2.0 | 0 |
| 312 | **Shepherd's pie**, vegetable | (2.6) | 2.4 | 0.8 | 0.8 | 0.7 | N | 0.08 | 0.4 | 1.3 | 1.9 | 0 |
| 313 | vegetable, retail | 1.5 | 1.2 | 0.4 | 0.6 | 0.2 | N | 0.02 | 2.0 | 1.2 | 1.4 | 7 |
| 314 | **Soya mince**, granules | 11.6 | N | 1.9 | N | N | 0.1 | 1.25 | (0.7) | (1.0) | (2.6) | 0 |
| 315 | **Sweet potato and onion layer** | 1.9 | 2.0 | 0.8 | 1.0 | 0.3 | 0.1 | 0.02 | 1.7 | 0.8 | 0.2 | 8 |
| 316 | **Tabouleh** | N | N | N | N | N | N | N | (0.4) | (1.5) | (2.0) | 0 |
| 317 | **Tagliatelle**, with vegetables, retail | 1.0 | 0.7 | 0.2 | 0.4 | 0.2 | 0.1 | 0.01 | 0.8 | 0.7 | 1.3 | 2 |
| 318 | **Tempeh burgers**, *fried in vegetable oil* | (2.2) | (1.8) | (0.4) | (0.6) | (0.7) | Tr | N | (0.9) | (2.3) | (2.9) | 23 |
| 319 | **Tofu burger**, *baked* | 2.2 | 2.2 | N | N | N | 0.2 | 0.49 | 0.6 | 1.0 | 1.9 | 0 |
| 320 | **Tofu spread** | 0.6 | 0.5 | 0.1 | 0.2 | Tr | Tr | 0.14 | 2.9 | 4.5 | 11.5 | 17 |

**Inorganic constituents per 100g food**

| No. 15- | Food | Na | K | Ca | Mg | P | Fe | Cu | Zn | Cl | Mn | Se | I |
|---|---|---|---|---|---|---|---|---|---|---|---|---|---|
| | | | | | | mg | | | | | | μg | |
| 303 | **Salad**, Waldorf | 88 | 130 | 13 | 12 | 35 | 0.3 | 0.10 | 0.3 | 150 | 0.3 | N | 7 |
| 304 | Waldorf, retail | 220 | 210 | 25 | 17 | 55 | 0.8 | 0.15 | 0.4 | 360 | 0.3 | N | (17) |
| 305 | **Samosas**, vegetable, retail | 390 | 150 | 65 | 19 | 65 | 1.5 | 0.11 | 0.5 | 590 | 0.3 | N | N |
| 306 | **Sauce**, curry, onion, *with butter* | 230 | 170 | 36 | 9 | 36 | 1.5 | 0.07 | 0.3 | 370 | 0.2 | (1) | (14) |
| 307 | *with vegetable oil* | 11 | 160 | 32 | 9 | 29 | 1.4 | 0.06 | 0.2 | 28 | 0.2 | (1) | (6) |
| 308 | curry, sweet | 16 | 160 | 27 | 11 | 25 | 1.5 | 0.08 | 0.2 | 32 | 0.2 | Tr | 2 |
| 309 | curry, tomato and onion | 400 | 330 | 27 | 19 | 37 | 0.9 | 0.09 | 0.3 | 660 | 0.2 | Tr | (6) |
| 310 | tomato base | 49 | 330 | 20 | 13 | 28 | 0.6 | 0.10 | 0.2 | 120 | 0.1 | Tr | (3) |
| 311 | tomato and mushroom | 39 | 350 | 25 | 14 | 35 | 0.8 | 0.11 | 0.2 | 110 | 0.2 | (1) | (4) |
| 312 | **Shepherd's pie**, vegetable | 81 | 330 | 16 | 22 | 75 | 1.7 | 0.23 | 0.8 | 160 | 0.3 | 17 | N |
| 313 | vegetable, retail | 340 | 240 | 12 | 14 | 36 | 0.6 | 0.07 | 0.3 | 560 | (0.1) | (7) | N |
| 314 | **Soya mince**, granules | 4420 | 2160 | 240 | 270 | 570 | 9.0 | 0.87 | 4.4 | 7300 | 3.1 | (4) | (17) |
| 315 | **Sweet potato and onion layer** | 170 | 300 | 93 | 40 | 96 | 0.7 | 0.12 | 0.5 | 270 | 0.3 | (2) | 9 |
| 316 | **Tabouleh** | 76 | 140 | 27 | 33 | 79 | 1.8 | N | N | 130 | N | N | N |
| 317 | **Tagliatelle**, with vegetables, retail | 6 | 100 | 13 | 8 | 29 | 0.4 | 0.04 | 0.2 | 22 | 0.1 | Tr | (1) |
| 318 | **Tempeh burgers**, *fried in vegetable oil* | 130 | 180 | 41 | 45 | 140 | 1.4 | 0.38 | 1.0 | N | 0.9 | N | N |
| 319 | **Tofu burger**, *baked* | 210 | 180 | 330 | 43 | 170 | 1.9 | 0.24 | 1.7 | 23 | 1.3 | N | N |
| 320 | **Tofu spread** | 690 | 120 | 250 | 17 | 66 | 1.0 | 0.11 | 0.4 | 760 | 0.2 | N | N |

| No. 15- | Food | Retinol µg | Carotene µg | Vitamin D µg | Vitamin E mg | Thiamin mg | Ribo-flavin mg | Niacin mg | Trypt 60 mg | Vitamin B6 mg | Vitamin B12 µg | Folate µg | Panto-thenate mg | Biotin µg | Vitamin C mg |
|---|---|---|---|---|---|---|---|---|---|---|---|---|---|---|---|
| 303 | **Salad,** Waldorf | 15 | 34 | 0.1 | 3.78 | 0.05 | 0.03 | 0.2 | 0.3 | 0.08 | 0.1 | 7 | (0.15) | N | 5 |
| 304 | Waldorf, retail | 38 | 59 | 0.2 | 8.91 | 0.07 | 0.05 | 0.2 | 0.4 | 0.11 | 0.2 | 13 | N | N | 5 |
| 305 | **Samosas,** vegetable, retail | 0 | N | 0 | N | 0.12 | 0.08 | 1.1 | 0.7 | 0.15 | 0 | 44 | N | N | N |
| 306 | **Sauce,** curry, onion, *with butter* | 240 | 140 | 0.2 | 0.83 | 0.08 | 0.01 | 0.5 | 0.3 | 0.12 | Tr | 6 | (0.07) | (0.5) | 4 |
| 307 | *with vegetable oil* | 0 | 13 | 0 | 7.40 | 0.08 | 0.01 | 0.5 | 0.3 | (0.12) | 0 | 6 | (0.07) | (0.5) | 4 |
| 308 | curry, sweet | 0 | 32 | 0 | 1.24 | 0.04 | 0.01 | 0.3 | 0.2 | 0.06 | 0 | 3 | 0.05 | 0.6 | 1 |
| 309 | curry, tomato and onion | 0 | 310 | 0 | 5.66 | 0.08 | 0.02 | 0.8 | 0.2 | 0.14 | 0 | 7 | 0.17 | 1.3 | 6 |
| 310 | tomato base | 0 | 320 | 0 | 1.73 | 0.06 | 0.02 | 0.8 | 0.2 | 0.13 | 0 | 7 | 0.21 | 1.5 | 7 |
| 311 | tomato and mushroom | 0 | 240 | 0 | 1.78 | 0.06 | 0.04 | 0.7 | 0.2 | 0.12 | 0 | 13 | 0.28 | 2.0 | 7 |
| 312 | **Shepherd's pie,** vegetable | 0 | 325 | 0 | 1.06 | 0.13 | 0.05 | 0.7 | 0.7 | 0.25 | 0 | 15 | (0.31) | (1.0) | 2 |
| 313 | vegetable, retail | 29 | 585 | Tr | (0.75) | 0.11 | 0.01 | 0.4 | 0.4 | 0.18 | Tr | 10 | 0.24 | (0.5) | 4 |
| 314 | **Soya mince,** granules | 0 | N | 0 | N | 0.54 | 0.42 | 1.8 | 8.2 | 0.51 | Tr | 35 | N | N | (1) |
| 315 | **Sweet potato and onion layer** | 29 | 2900 | Tr | 3.31 | 0.08 | 0.07 | 0.5 | 0.8 | 0.08 | 0.2 | 6 | 0.45 | N | 7 |
| 316 | **Tabouleh** | 0 | 260 | 0 | (1.15) | 0.12 | 0.04 | 1.1 | 0.4 | N | 0 | 17 | N | N | 13 |
| 317 | **Tagliatelle,** with vegetables, retail | 10 | 170 | Tr | 0.63 | 0.03 | 0.01 | 0.3 | 0.3 | 0.05 | Tr | 4 | 0.07 | 0.4 | 2 |
| 318 | **Tempeh burgers,** *fried in vegetable oil* | 11 | 8 | 0.1 | 1.55 | 0.13 | 0.16 | 1.7 | 1.8 | N | 0.2 | 13 | N | N | Tr |
| 319 | **Tofu burger,** *baked* | 0 | 655 | 0 | 1.84 | 0.22 | 0.05 | 0.5 | 1.5 | 0.22 | 0 | 17 | 0.25 | N | 1 |
| 320 | **Tofu spread** | 20 | 270 | 0.1 | 5.05 | 0.06 | 0.04 | 0.4 | 0.8 | 0.06 | 0.1 | 13 | 0.08 | N | 3 |

No.   Food
**15-**

Description and main data sources

| No. | Food | Description and main data sources |
|---|---|---|
| 321 | **Tomatoes,** stuffed with rice | Greek dish. Recipe from a personal collection |
| 322 | stuffed with vegetables | Sweetcorn, onion and mushroom stuffing. Recipe from review of recipe collection |
| 323 | **Vegebanger mix** | 8 samples, 2 brands; assorted flavours |
| 324 | *made up with water* | Calculated from No 323 made up with water according to packet directions |
| 325 | *made up with water, fried in sunflower oil* | Calculated from No 324 using 15% weight loss and measured fat uptake |
| 326 | *made up with water, fried in vegetable oil* | Calculated from No 324 using 15% weight loss and measured fat uptake |
| 327 | *made up with water and egg* | Calculated from No 323 made up with water and egg according to packet directions |
| 328 | *made up with water and egg, fried in sunflower oil* | Calculated from No 327 using 15% weight loss and measured fat uptake |
| 329 | *made up with water and egg, fried in vegetable oil* | Calculated from No 327 using 15% weight loss and measured fat uptake |
| 330 | **Vegeburger,** retail, *fried in vegetable oil* | 10 samples, 5 brands; soya and wheat protein based. Fried 4-5 minutes |
| 331 | *grilled* | 6 samples, 3 brands; soya protein based. Grilled 6-10 minutes |
| 332 | **Vegeburger mix** | 10 samples, 6 brands |
| 333 | *made up with water* | Calculated from No 332 made up with water according to packet directions |
| 334 | *made up with water, fried in sunflower oil* | Calculated from No 333 using 12% weight loss and measured fat uptake |

| No. 15- | Food | Water g | Total nitrogen g | Protein g | Fat g | Carbohydrate g | Energy value kcal | kJ |
|---|---|---|---|---|---|---|---|---|
| 321 | **Tomatoes**, stuffed with rice | 60.1 | 0.36 | 2.1 | 13.4 | 22.2 | 212 | 887 |
| 322 | stuffed with vegetables | 74.9 | 0.40 | 2.2 | 6.2 | 12.8 | 111 | 469 |
| 323 | **Vegebanger mix** | 5.9 | 5.90 | 36.9 | 15.3 | 28.8 | 393 | 1654 |
| 324 | *made up with water* | 64.5 | 2.22 | 13.9 | 5.8 | 10.9 | 148 | 623 |
| 325 | *made up with water, fried in sunflower oil* | 52.2 | 2.41 | 15.1 | 15.6 | 11.8 | 244 | 1022 |
| 326 | *made up with water, fried in vegetable oil* | 52.2 | 2.41 | 15.1 | 15.6 | 11.8 | 244 | 1022 |
| 327 | *made up with water and egg* | 58.6 | 2.68 | 16.7 | 7.9 | 11.3 | 180 | 759 |
| 328 | *made up with water and egg, fried in sunflower oil* | 45.6 | 2.89 | 18.1 | 18.2 | 12.2 | 281 | 1176 |
| 329 | *made up with water and egg, fried in vegetable oil* | 45.6 | 2.89 | 18.1 | 18.2 | 12.2 | 281 | 1176 |
| 330 | **Vegeburger**, retail, fried in vegetable oil | 49.6 | 2.78 | 15.9 | 17.0 | 7.5 | 245 | 1019 |
| 331 | grilled | 50.3 | 2.91 | 16.6 | 11.1 | 8.0 | 196 | 821 |
| 332 | **Vegeburger mix,** | 6.1 | 4.86 | 30.4 | 12.6 | 34.5 | 364 | 1535 |
| 333 | *made up with water* | 64.6 | 1.83 | 11.5 | 4.8 | 13.0 | 137 | 578 |
| 334 | *made up with water, fried in sunflower oil* | 55.6 | 1.96 | 12.3 | 11.6 | 13.9 | 205 | 860 |

Carbohydrate fractions, g per 100g food

| No. Food | Starch | Total sugars | Individual sugars | | | | | | | Oligo- saccharides |
| | | | Glucose | Fructose | Galactose | Sucrose | Maltose | Lactose | |
|---|---|---|---|---|---|---|---|---|---|---|
| 15- | | | | | | | | | | |
| 321 **Tomatoes**, stuffed with rice | 14.4 | 7.7 | (2.9) | (2.9) | 0 | (1.7) | Tr | 0 | 0.2 |
| 322 stuffed with vegetables | 5.0 | 7.4 | 2.2 | 2.2 | 0 | 2.9 | Tr | 0 | 0.4 |
| 323 **Vegebanger mix** | 25.5 | 3.3 | 0.4 | 0.6 | 0 | 2.3 | Tr | 0 | 0 |
| 324 made up with water | 9.6 | 1.2 | 0.2 | 0.2 | 0 | 0.9 | Tr | 0 | 0 |
| 325 made up with water, fried in sunflower oil | 10.4 | 1.4 | 0.2 | 0.2 | 0 | 0.9 | Tr | 0 | 0 |
| 326 made up with water, fried in vegetable oil | 10.4 | 1.4 | 0.2 | 0.2 | 0 | 0.9 | Tr | 0 | 0 |
| 327 made up with water and egg | 10.0 | 1.3 | 0.2 | 0.2 | 0 | 0.9 | Tr | 0 | 0 |
| 328 made up with water and egg, fried in sunflower oil | 10.8 | 1.4 | 0.2 | 0.3 | 0 | 1.0 | Tr | 0 | 0 |
| 329 made up with water and egg, fried in vegetable oil | 10.8 | 1.4 | 0.2 | 0.3 | 0 | 1.0 | Tr | 0 | 0 |
| 330 **Vegeburger**, retail, fried in vegetable oil | 4.1 | 3.4 | 0.5 | 0.4 | 0 | 1.7 | 0.8 | 0 | 0 |
| 331 grilled | 4.4 | 3.6 | 0.5 | 0.4 | 0 | 1.8 | 0.9 | 0 | 0 |
| 332 **Vegeburger mix** | 29.7 | 4.8 | 0.4 | 0.5 | 0 | 2.5 | 1.4 | 0 | 0 |
| 333 made up with water | 11.2 | 1.8 | 0.2 | 0.2 | 0 | 0.9 | 0.5 | 0 | 0 |
| 334 made up with water, fried in sunflower oil | 12.0 | 1.9 | 0.2 | 0.2 | 0 | 1.0 | 0.6 | 0 | 0 |

**Fibre fractions, phytic acid and fatty acids, g per 100g food**
**Cholesterol, mg per 100g food**

| No. 15- | Food | Dietary fibre, g Southgate method | Dietary fibre, g Englyst method | Fibre fractions, g Cellulose | Non-cellulosic polysaccharide Soluble | Non-cellulosic polysaccharide Insoluble | Lignin | Phytic acid g | Fatty acids, g Satd | Mono-unsatd | Poly-unsatd | Cholesterol mg |
|---|---|---|---|---|---|---|---|---|---|---|---|---|
| 321 | **Tomatoes**, stuffed with rice | 2.0 | 1.1 | 0.4 | 0.5 | 0.3 | (0.4) | (0.08) | 2.0 | 9.0 | 1.8 | 0 |
| 322 | stuffed with vegetables | 3.2 | 1.9 | 0.7 | 0.6 | 0.6 | 0.4 | 0.05 | 0.7 | 2.1 | 3.0 | 0 |
| 323 | **Vegebanger mix** | N | 8.1 | 1.6 | 3.1 | 3.4 | N | N | 6.2 | 2.8 | 3.3 | 0 |
| 324 | made up with water | N | 3.1 | 0.6 | 1.2 | 1.3 | N | N | 2.3 | 1.1 | 1.2 | 0 |
| 325 | made up with water, fried in sunflower oil | | 3.3 | 0.7 | 1.3 | 1.4 | N | N | 3.7 | 3.0 | 7.3 | 0 |
| 326 | made up with water, fried in vegetable oil | N | 3.3 | 0.7 | 1.3 | 1.4 | N | N | 3.5 | 4.5 | 5.9 | 0 |
| 327 | made up with water and egg | N | 3.2 | 0.6 | 1.2 | 1.3 | N | N | 3.0 | 1.9 | 1.5 | 68 |
| 328 | made up with water and egg, fried in sunflower oil | N | 3.4 | 0.7 | 1.3 | 1.4 | N | N | 4.4 | 4.0 | 7.7 | 74 |
| 329 | made up with water and egg, fried in vegetable oil | N | 3.4 | 0.7 | 1.3 | 1.4 | N | N | 4.2 | 5.5 | 6.3 | 74 |
| 330 | **Vegeburger**, retail, fried in vegetable oil | N | 3.6 | 0.7 | 1.7 | 1.2 | N | N | N | N | N | N |
| 331 | grilled | N | 4.2 | 0.8 | 2.0 | 1.4 | N | N | N | N | N | N |
| 332 | **Vegeburger mix** | N | 8.6 | 1.3 | 3.6 | 3.7 | N | N | 4.3 | 4.5 | 1.2 | 0 |
| 333 | made up with water | N | 3.2 | 0.5 | 1.4 | 1.4 | N | N | 1.6 | 1.7 | 0.5 | 0 |
| 334 | made up with water, fried in sunflower oil | N | 3.5 | 0.5 | 1.5 | 1.5 | N | N | 2.5 | 3.1 | 4.6 | 0 |

# Vegetable Dishes

**Inorganic constituents per 100g food**

**15-321 to 15-334**

| No. 15- | Food | Na | K | Ca | Mg | P | Fe | Cu | Zn | Cl | Mn | Se | I |
|---|---|---|---|---|---|---|---|---|---|---|---|---|---|
| | | | | | | mg | | | | | | µg | |
| 321 | **Tomatoes**, stuffed with rice | 89 | 280 | 24 | 13 | 50 | 0.7 | 0.11 | 0.4 | 170 | 0.3 | (3) | (4) |
| 322 | stuffed with vegetables | 310 | 440 | 16 | 19 | 68 | 0.9 | 0.14 | 0.4 | 520 | 0.2 | 1 | N |
| 323 | **Vegebanger mix** | 1310 | 970 | 150 | 130 | 360 | 7.3 [a] | 0.54 | 2.8 | 1720 | 2.2 | 21 | N |
| 324 | made up with water | 490 | 370 | 56 | 49 | 140 | 2.8 | 0.20 | 1.1 | 650 | 0.8 | 8 | N |
| 325 | made up with water, fried in sunflower oil | 530 | 400 | 61 | 53 | 150 | 3.0 | 0.22 | 1.1 | 700 | 0.9 | 9 | N |
| 326 | made up with water, fried in vegetable oil | 530 | 400 | 61 | 53 | 150 | 3.0 | 0.22 | 1.1 | 700 | 0.9 | 9 | N |
| 327 | made up with water and egg | 540 | 400 | 69 | 53 | 180 | 3.2 | 0.23 | 1.3 | 700 | 0.9 | 10 | N |
| 328 | made up with water and egg, fried in sunflower oil | 580 | 440 | 74 | 57 | 190 | 3.5 | 0.24 | 1.4 | 760 | 0.9 | 11 | N |
| 329 | made up with water and egg, fried in vegetable oil | 580 | 440 | 74 | 57 | 190 | 3.5 | 0.24 | 1.4 | 760 | 0.9 | 11 | N |
| 330 | **Vegeburger**, retail, fried in vegetable oil | 420 | 530 | 90 | 69 | 210 | 3.9 [b] | 0.35 | 1.4 | 570 | 1.0 | 7 | N |
| 331 | grilled | 490 | 610 | 100 | 80 | 240 | 4.5 [b] | 0.40 | 1.6 | 660 | 1.1 | 8 | N |
| 332 | **Vegeburger mix** | 1050 | 1020 | 240 | 150 | 430 | 8.9 [a] | 0.59 | 3.1 | 1480 | 2.5 | 11 | N |
| 333 | made up with water | 400 | 380 | 90 | 56 | 160 | 3.4 | 0.22 | 1.2 | 560 | 0.9 | 4 | N |
| 334 | made up with water, fried in sunflower oil | 420 | 410 | 96 | 60 | 170 | 3.6 | 0.24 | 1.3 | 600 | 1.0 | 4 | N |

[a] 3 samples contained added Fe

[b] 1 sample contained added Fe

# Vegetable Dishes

| No. Food 15- | Retinol µg | Carotene µg | Vitamin D µg | Vitamin E mg | Thiamin mg | Ribo-flavin mg | Niacin mg | Trypt 60 mg | Vitamin B6 mg | Vitamin B12 µg | Folate µg | Panto-thenate mg | Biotin µg | Vitamin C mg |
|---|---|---|---|---|---|---|---|---|---|---|---|---|---|---|
| 321 Tomatoes, stuffed with rice | 0 | 545 | 0 | (1.66) | 0.11 | 0.01 | 1.2 | 0.4 | 0.14 | 0 | 8 | (0.23) | (1.4) | 7 |
| 322 stuffed with vegetables | 0 | 775 | 0 | 2.97 | 0.12 | 0.06 | 1.8 | 0.4 | 0.21 | 0 | 15 | 0.54 | N | 10 |
| 323 Vegebanger mix | 0 | N | 0 | N | 0.41 a | 0.37 a | 2.3 | 7.9 | 0.50 | N | 110 | (1.60) | (16.0) | Tr |
| 324 made up with water | 0 | N | 0 | N | 0.15 | 0.14 | 0.9 | 3.0 | 0.19 | N | 41 | (0.60) | (6.0) | Tr |
| 325 made up with water, fried in sunflower oil | 0 | N | 0 | N | 0.13 | 0.15 | 0.9 | 3.2 | 0.15 | N | 20 | (0.65) | (6.5) | Tr |
| 326 made up with water, fried in vegetable oil | 0 | N | 0 | N | 0.13 | 0.15 | 0.9 | 3.2 | 0.15 | N | 20 | (0.65) | (6.5) | Tr |
| 327 made up with water and egg | 33 | N | 0.3 | N | 0.18 | 0.23 | 0.9 | 3.8 | 0.22 | N | 52 | (0.94) | (9.9) | Tr |
| 328 made up with water and egg, fried in sunflower oil | 36 | N | 0.3 | N | 0.15 | 0.25 | 1.0 | 4.1 | 0.18 | N | 25 | (1.02) | (10.6) | Tr |
| 329 made up with water and egg, fried in vegetable oil | 36 | N | 0.3 | N | 0.15 | 0.25 | 1.0 | 4.1 | 0.18 | N | 25 | (1.02) | (10.6) | Tr |
| 330 Vegeburger, retail, fried in vegetable oil | N | Tr | N | N | 2.08 b | 0.36 b | 2.4 c | 3.7 | 0.26 | N c | 82 | N | N | N |
| 331 grilled | N | Tr | N | N | 2.40 d | 0.42 d | 2.8 e | 3.9 | 0.30 e | N | 95 | N | N | N |
| 332 Vegeburger mix | 0 | N | 0 | N | 2.20 f | 0.81 f | 3.9 | 6.5 | 0.96 | N | 110 | 1.60 | 16.0 | Tr |
| 333 made up with water | 0 | N | 0 | N | 0.83 | 0.31 | 1.5 | 2.5 | 0.36 | N | 41 | 0.60 | 6.0 | Tr |
| 334 made up with water, fried in sunflower oil | 0 | N | 0 | N | 0.71 | 0.33 | 1.6 | 2.6 | 0.29 | N | 19 | 0.65 | 6.5 | Tr |

a  3 samples contained added thiamin and riboflavin
c  3 samples contained added niacin and vitamin B12
e  1 sample contained added niacin and vitamin B6

b  4 samples contained added thiamin and riboflavin
d  2 samples contained added thiamin and riboflavin
f  3 samples contained added thiamin and riboflavin

| No. Food 15- | Description and main data sources |
|---|---|
| 335 **Vegeburger mix**, *made up with water, fried in vegetable oil* | Calculated from No 333 using 12% weight loss and measured fat uptake |
| 336 *made up with water, grilled* | Calculated from No 333 using 9% weight loss |
| 337 *made up with water and egg* | Calculated from No 332 made up with water and egg according to packet directions |
| 338 *made up with water and egg, fried in sunflower oil* | Calculated from No 337 using 12% weight loss and measured fat uptake |
| 339 *made up with water and egg, fried in vegetable oil* | Calculated from No 337 using 12% weight loss and measured fat uptake |
| 340 *made up with water and egg, grilled* | Calculated from No 337 using 9% weight loss |
| 341 **Vegetable bake** | Assorted vegetables topped with cheese sauce and breadcrumbs. Recipe from dietary survey records |
| 342 **Vegetable pancake roll** | 18 samples purchased from take-away outlets; Chinese pancake rolls |
| 343 **Vegetable pate** | 5 samples, 4 brands; assorted soya, cereal and vegetable based |
| 344 **Vegetable stir fry mix**, *fried in corn oil* | 8 assorted frozen samples; stir-fried 4-7 minutes |
| 345 *fried in sunflower oil* | 8 assorted frozen samples; stir fried 4-7 minutes |
| 346 *fried in vegetable oil* | 8 assorted frozen samples; stir-fried 4-7 minutes |
| 347 **Vine leaves**, stuffed with rice | Greek dish. Recipe from a personal collection |

| No. Food | Water | Total nitrogen | Protein | Fat | Carbohydrate | Energy value | |
|---|---|---|---|---|---|---|---|
| 15- | g | g | g | g | g | kcal | kJ |
| 335 **Vegeburger mix**, *made up with* | | | | | | | |
| *water, fried in vegetable oil* | 55.6 | 1.96 | 12.3 | 11.6 | 13.9 | 205 | 860 |
| 336 *made up with water, grilled* | 61.1 | 2.01 | 12.6 | 5.2 | 14.3 | 150 | 635 |
| 337 *made up with water and egg* | 63.4 | 1.99 | 12.5 | 6.0 | 12.1 | 148 | 626 |
| 338 *made up with water and* | | | | | | | |
| *egg, fried in sunflower oil* | 54.3 | 2.14 | 13.4 | 12.9 | 13.0 | 218 | 912 |
| 339 *made up with water and* | | | | | | | |
| *egg, fried in vegetable oil* | 54.3 | 2.14 | 13.4 | 12.9 | 13.0 | 218 | 912 |
| 340 *made up with water and* | | | | | | | |
| *egg, grilled* | 59.8 | 2.19 | 13.7 | 6.6 | 13.3 | 163 | 688 |
| 341 **Vegetable bake** | 74.7 | 0.67 | 4.2 | 7.1 | 12.4 | 127 | 532 |
| 342 **Vegetable pancake roll** | 58.3 | 1.06 | 6.6 | 12.5 | 21.0 | 218 | 911 |
| 343 **Vegetable pate** | 64.9 | 1.20 | 7.5 | 13.4 | 5.9 | 173 | 718 |
| 344 **Vegetable stir fry mix**, *fried in* | | | | | | | |
| *corn oil* | 83.8 | 0.32 | 2.0 | 3.6 | 6.4 | 64 | 270 |
| 345 *fried in sunflower oil* | 83.8 | 0.32 | 2.0 | 3.6 | 6.4 | 64 | 270 |
| 346 *fried in vegetable oil* | 83.8 | 0.32 | 2.0 | 3.6 | 6.4 | 64 | 270 |
| 347 **Vine leaves**, stuffed with rice | 48.5 | 0.46 | 2.8 | 18.0 | 23.8 | 262 | 1094 |

## Vegetable Dishes

**15-335 to 15-347**

### Carbohydrate fractions, g per 100g food

| No. 15- | Food | Starch | Total sugars | Individual sugars | | | | | | Oligo-saccharides |
|---|---|---|---|---|---|---|---|---|---|---|
| | | | | Glucose | Fructose | Galactose | Sucrose | Maltose | Lactose | |
| 335 | **Vegeburger mix**, made up with water, fried in vegetable oil | 12.0 | 1.9 | 0.2 | 0.2 | 0 | 1.0 | 0.6 | 0 | 0 |
| 336 | made up with water, grilled | 12.3 | 2.0 | 0.2 | 0.2 | 0 | 1.0 | 0.6 | 0 | 0 |
| 337 | made up with water and egg | 10.5 | 1.7 | 0.1 | 0.2 | 0 | 0.9 | 0.5 | 0 | 0 |
| 338 | made up with water and egg, fried in sunflower oil | 11.2 | 1.8 | 0.2 | 0.2 | 0 | 0.9 | 0.5 | 0 | 0 |
| 339 | made up with water and egg, fried in vegetable oil | 11.2 | 1.8 | 0.2 | 0.2 | 0 | 0.9 | 0.5 | 0 | 0 |
| 340 | made up with water and egg, grilled | 11.5 | 1.9 | 0.2 | 0.2 | 0 | 1.0 | 0.5 | 0 | 0 |
| 341 | **Vegetable bake** | 7.6 | 4.5 | 0.8 | 0.7 | 0 | 0.9 | Tr | 2.0 | 0.3 |
| 342 | **Vegetable pancake roll** | 18.0 | 3.0 | (0.8) | (0.8) | 0 | 1.4 | Tr | 0 | Tr |
| 343 | **Vegetable pate** | 5.7 | 0.2 | Tr | 0.2 | 0 | Tr | Tr | 0 | 0 |
| 344 | **Vegetable stir fry mix**, fried in corn oil | 2.5 | 3.9 | 1.4 | 1.2 | 0 | 1.1 | 0.2 | 0 | Tr |
| 345 | fried in sunflower oil | 2.5 | 3.9 | 1.4 | 1.2 | 0 | 1.1 | 0.2 | 0 | Tr |
| 346 | fried in vegetable oil | 2.5 | 3.9 | 1.4 | 1.2 | 0 | 1.1 | 0.2 | 0 | Tr |
| 347 | **Vine leaves**, stuffed with rice | 12.3 | 10.5 | 4.9 | 4.6 | 0 | 0.9 | 0 | 0 | 1.1 |

**Fibre fractions, phytic acid and fatty acids, g per 100g food**
**Cholesterol, mg per 100g food**

| No. Food 15- | Dietary fibre, g Southgate method | Englyst method | Fibre fractions, g Cellulose | Non-cellulosic polysaccharide Soluble | Insoluble | Lignin | Phytic acid g | Fatty acids, g Satd | Mono- unsatd | Poly- unsatd | Cholesterol mg |
|---|---|---|---|---|---|---|---|---|---|---|---|
| 335 **Vegeburger mix**, *made up with water, fried in vegetable oil* | N | 3.5 | 0.5 | 1.5 | 1.5 | N | N | 2.4 | 4.1 | 3.6 | 0 |
| 336 *made up with water, grilled* | N | 3.6 | 0.5 | 1.5 | 1.5 | N | N | 1.8 | 1.9 | 0.5 | 0 |
| 337 *made up with water and egg* | N | 3.0 | 0.5 | 1.3 | 1.3 | N | N | 2.0 | 2.3 | 0.6 | 54 |
| 338 *made up with water and egg, fried in sunflower oil* | N | 3.2 | 0.5 | 1.4 | 1.4 | N | N | 2.9 | 3.7 | 4.8 | 58 |
| 339 *made up with water and egg, fried in vegetable oil* | N | 3.2 | 0.5 | 1.4 | 1.4 | N | N | 2.8 | 4.7 | 3.8 | 58 |
| 340 *made up with water and egg, grilled* | N | 3.3 | 0.5 | 1.4 | 1.4 | N | N | 2.1 | 2.5 | 0.7 | 59 |
| 341 **Vegetable bake** | (1.4) | 1.1 | 0.3 | 0.6 | 0.2 | 0.1 | 0.02 | 3.0 | 2.1 | 1.5 | 11 |
| 342 **Vegetable pancake roll** | N | N | N | N | N | N | N | 3.5 | 6.0 | 2.5 | 0 |
| 343 **Vegetable pate** | N | N | N | N | N | Tr | N | N | N | N | Tr |
| 344 **Vegetable stir fry mix**, *fried in corn oil* | N | N | N | N | N | N | N | 0.5 | 0.9 | 2.1 | 0 |
| 345 *fried in sunflower oil* | N | N | N | N | N | N | N | 0.4 | 0.7 | 2.3 | 0 |
| 346 *fried in vegetable oil* | N | N | N | N | N | N | N | 0.3 | 1.8 | 1.3 | 0 |
| 347 **Vine leaves**, stuffed with rice | 2.9 | N | N | N | N | 0.1 | 0.07 | 2.6 | 12.3 | 2.2 | 0 |

| No. 15- | Food | Na | K | Ca | Mg | P | mg Fe | Cu | Zn | Cl | Mn | µg Se | I |
|---|---|---|---|---|---|---|---|---|---|---|---|---|---|
| 335 | **Vegeburger mix**, *made up with water, fried in vegetable oil* | 420 | 410 | 96 | 60 | 170 | 3.6 | 0.24 | 1.3 | 600 | 1.0 | 4 | N |
| 336 | *made up with water, grilled* | 430 | 420 | 99 | 62 | 180 | 3.7 | 0.24 | 1.3 | 610 | 1.0 | 5 | N |
| 337 | *made up with water and egg* | 390 | 380 | 92 | 54 | 180 | 3.4 | 0.22 | 1.3 | 540 | 0.9 | 5 | N |
| 338 | *made up with water and egg, fried in sunflower oil* | 420 | 400 | 99 | 58 | 190 | 3.6 | 0.23 | 1.4 | 580 | 0.9 | 6 | N |
| 339 | *made up with water and egg, fried in vegetable oil* | 420 | 400 | 99 | 58 | 190 | 3.6 | 0.23 | 1.4 | 580 | 0.9 | 6 | N |
| 340 | *made up with water and egg, grilled* | 430 | 410 | 100 | 59 | 200 | 3.7 | 0.24 | 1.4 | 600 | 1.0 | 6 | N |
| 341 | **Vegetable bake** | 120 | 230 | 100 | 14 | 89 | 0.5 | 0.04 | 0.5 | 210 | 0.1 | (3) | (12) |
| 342 | **Vegetable pancake roll** | 610 | N | N | N | N | 2.1 | N | N | 950 | N | N | N |
| 343 | **Vegetable pate** | 540 | 300 | 130 | 28 | 190 | 4.2 | 0.13 | 2.1 | 810 | 0.5 | N | N |
| 344 | **Vegetable stir fry mix**, *fried in corn oil* | 11 | 230 | 30 | 16 | 46 | 0.5 | 0.11 | 0.3 | 27 | 0.1 | Tr | N |
| 345 | *fried in sunflower oil* | 11 | 230 | 30 | 16 | 46 | 0.5 | 0.11 | 0.3 | 27 | 0.1 | Tr | N |
| 346 | *fried in vegetable oil* | 11 | 230 | 30 | 16 | 46 | 0.5 | 0.11 | 0.3 | 27 | 0.1 | Tr | N |
| 347 | **Vine leaves**, stuffed with rice | 1140 | 190 | 130 | 24 | 53 | 1.1 | 0.56 | 0.5 | 1790 | 0.4 | (2) | (4) |

| No. 15- | Food | Retinol µg | Carotene µg | Vitamin D µg | Vitamin E mg | Thiamin mg | Riboflavin mg | Niacin mg | Trypt 60 mg | Vitamin B6 mg | Vitamin B12 µg | Folate µg | Pantothenate mg | Biotin µg | Vitamin C mg |
|---|---|---|---|---|---|---|---|---|---|---|---|---|---|---|---|
| 335 | **Vegeburger mix**, *made up with water, fried in vegetable oil* | 0 | N | 0 | N | 0.71 | 0.33 | 1.6 | 2.6 | 0.29 | N | 19 | 0.65 | 6.5 | Tr |
| 336 | *made up with water, grilled* | 0 | N | 0 | N | 0.73 | 0.27 | 1.3 | 2.7 | 0.32 | N | 22 | 0.53 | 5.3 | Tr |
| 337 | *made up with water and egg* | 26 | N | 0.3 | N | 0.79 | 0.35 | 1.4 | 2.8 | 0.35 | N | 45 | 0.81 | 8.5 | Tr |
| 338 | *made up with water and egg, fried in sunflower oil* | 28 | N | 0.3 | N | 0.67 | 0.38 | 1.5 | 3.0 | 0.29 | N | 22 | 0.87 | 9.1 | Tr |
| 339 | *made up with water and egg, fried in vegetable oil* | 28 | N | 0.3 | N | 0.67 | 0.38 | 1.5 | 3.0 | 0.29 | N | 22 | 0.87 | 9.1 | Tr |
| 340 | *made up with water and egg, grilled* | 29 | N | 0.3 | N | 0.69 | 0.31 | 1.2 | 3.1 | 0.31 | N | 25 | 0.71 | 7.4 | Tr |
| 341 | **Vegetable bake** | 72 | 1685 | 0.4 | 0.22 | 0.09 | 0.08 | 0.4 | 1.0 | 0.15 | 0.2 | 10 | 0.26 | (1.0) | 3 |
| 342 | **Vegetable pancake roll** | 0 | 3 | 0 | 0.77 | 0.09 | 0.05 | N | 1.2 | N | 0 | N | N | N | Tr |
| 343 | **Vegetable pate** | Tr a | N | Tr a | N | 2.10 | 1.30 | 4.7 | 1.6 | 0.46 | Tr a | 110 | N | N | Tr |
| 344 | **Vegetable stir fry mix**, *fried in corn oil* | 0 | N | 0 | 0 | 0.07 | 0.13 | 1.0 | 0.3 | 0.25 | 0 | 16 | N | N | 8 |
| 345 | *fried in sunflower oil* | 0 | N | 0 | N | 0.07 | 0.13 | 1.0 | 0.3 | 0.25 | 0 | 16 | N | N | 8 |
| 346 | *fried in vegetable oil* | 0 | N | 0 | N | 0.07 | 0.13 | 1.0 | 0.3 | 0.25 | 0 | 16 | N | N | 8 |
| 347 | **Vine leaves**, *stuffed with rice* | 0 | (420) | 0 | (1.10) | 0.13 | 0.06 | 0.9 | N | (0.13) | 0 | (6) | (0.12) | (1.1) | 4 |

a From egg present in one sample

# Appendices

# RECORDS

---

- For all the recipes, ingredient quantities are reported exactly as in the original recipe source or sources. Quantities have not been included for recipes obtained in confidence from manufacturers. The ingredients have however, been listed in quantity order.

- Where a recipe source indicated a portion but not quantity of an ingredient the portion size was taken from Crawley (1990) or weighed during recipe testing.

- In the recipe calculations, an egg has been assumed to weigh 50g. A level teaspoon refers to a standard 5ml spoon and has been taken to hold 9g marmite, 5g salt or sugar, 3.5g baking powder or bicarbonate of soda and 3g spices.

- Where present in recipes, lemon juice, milk, vinegar and water have been entered in millilitres. For milks the millilitre measures have been converted to gram weights for the purpose of recipe calculation.

- The type of fat used in the recipes has been specified. The vegetable oil was a blended vegetable oil. Margarine was an average of soft vegetable based and polyunsaturated only. The butter was salted.

- Unless specified the recipes use whole pasteurised milk, pasteurised single cream, Cheddar cheese and plain white flour. Sauces were made with whole milk unless specified. For recipes containing cabbage, the cabbage values used were for an average of winter, summer and white varieties.

- The baking powder used was a proprietary preparation whose composition is given in the Fifth Edition of McCance and Widdowson's *The Composition of Foods*. Use of another brand could result in a different composition in the cooked dish with respect to sodium, calcium and phosphorus.

- Where beans are given as an ingredient the profile was calculated from boiled and canned blackeye, butter, red kidney and soya beans.

- Where canned fruit are used as ingredients, the nutrient profile is an average of the fruit canned in syrup and in juice.

- Wherever possible, a measure of water absorbed both on soaking and boiling of beans has been given.

- For fried dishes the fat absorbed during frying has been included at the end of the ingredients list with the quantity absorbed shown in brackets.

- For a number of recipes obtained from dietary survey records only the major ingredients were recorded. These recipes do not contain a measure for salt, spices or other 'lesser' ingredients and these were not therefore, included in the recipe calculation.

# Recipes of ingredients used within the main recipes

## Cheese sauce

350ml whole, semi-skimmed or ½tsp salt
  skimmed milk             25g flour
75g cheese                 25g margarine

Melt the fat in a pan, add flour and cook gently for a few minutes stirring all the time. Add milk and cook until mixture thickens, stirring continually. Add grated cheese and seasoning. Reheat to soften the cheese, serve immediately.

Weight loss: 15%

## Flaky pastry

200g flour                85ml water
150g margarine         10ml lemon juice
½tsp salt

Divide margarine into four portions. Sift flour and salt, rub in one portion of fat. Mix with water and lemon juice, then knead until smooth and leave for 15 mins. Roll out, dot two-thirds with another fat portion and fold into 3. Roll out and repeat process with remaining 2 fat portions. Bake at 220°C/mark 7.

Weight loss: 24%

## French dressing

25ml vinegar            ½tsp salt
75g sunflower oil       ½tsp pepper

Shake the ingredients together in a screw-topped jar or bottle.

## Garam masala

50g coriander seeds, ground    10g cloves, ground
50g pepper, ground           10g cardamom, ground
40g cumin seeds, ground

## Mixed herbs

25g marjoram            25g sage
25g parsley              25g thyme

## Mixed nuts

67% peanuts            8% cashews
17% almonds           7% hazelnuts

## Mixed spices

25g curry powder
25g garam masala

25g paprika
25g chilli powder

## Pancakes

112g white or wholemeal flour
300ml whole milk
1 egg

56g margarine (for pan)
¼tsp salt

Sieve the flour into a basin, add the egg and about 100ml of the milk, stirring until smooth. Add the rest of the milk and beat to a smooth batter. Heat a little of the margarine in a frying pan and pour enough batter to cover the bottom. Cook both sides and turn out. Repeat until all the batter is used.

Weight loss: 20%

## Pizza dough

200g white or wholemeal flour
1 tsp salt
1 tsp sugar

150ml warm water
15g fresh or 2 tsp dried yeast

Mix all ingredients together to form a dough.

## Shortcrust pastry

200g white or wholemeal flour
100g margarine

½tsp salt
30ml water

Rub the fat into the flour, mix to a stiff dough with the water, roll out and bake at 200°C/mark 6.

Weight loss: 14%

## White sauce

350ml whole, semi-skimmed or
 skimmed milk
25g flour

½tsp salt
25g margarine

Melt margarine in a pan. Add flour and cook for a few minutes, stirring constantly. Add milk and salt and cook gently until mixture thickens.

Weight loss: 18%

## Main Recipes

### 1  Aubergine, stuffed with lentils and vegetables

| | |
|---|---|
| 55% cooked aubergine | 6.3% celery |
| 12.7% boiled whole lentils | 4.4% vegetable oil |
| 12.7% tomatoes | 1.3% soya sauce |
| 7.6% mushrooms | |

Proportions are derived from dietary survey records.

### 2  Aubergine, stuffed with rice

| | |
|---|---|
| 520g aubergines | 30g raisins |
| 40g chopped onion | 170g boiled white rice |
| 10g vegetable oil | 5g chopped fresh parsley |
| 85g chopped and skinned tomatoes | ½tsp salt |

Prick aubergines, cut in half and bake, flesh side down for 30 minutes at 190°C/mark 5. Fry the onions, remove from heat and mix with remaining ingredients. Scoop flesh from cooked aubergines and mix with rice mixture. Pile filling into aubergines, place under a grill until heated through, or back in oven at 190°C/mark 5 for approximately 15 minutes.

Weight loss: 29%

### 3  Aubergine, stuffed with vegetables, cheese topping

| | |
|---|---|
| 520g aubergines | 1tsp mixed herbs |
| 30g vegetable oil | ½tsp salt |
| 90g chopped onion | ¼tsp pepper |
| 150g sliced mushrooms | 35g tomato purée |
| 200g peeled tomatoes | 100g grated cheese |

Prick aubergines, cut in half and bake, flesh side down for 30 minutes at 190°C/mark 5. Fry the onions and mushrooms, add tomatoes, herbs, salt and pepper and reduce. Scoop flesh from cooked aubergines and mix with tomato mixture. Re-fill aubergine cases, sprinkle with cheese and place under a grill or bake in oven at 190°C/mark 5 for 15 minutes.

Weight loss: 38%

### 4  Bean loaf

| | |
|---|---|
| 90g chopped onion | 20g wholemeal four |
| 50g chopped mushrooms | 50g tomatoes |
| 20g vegetable oil | 20g tomato purée |
| | 210g beans |

Fry the onion and mushrooms in the oil, add flour, tomatoes and tomato purée. Mix with the mashed beans. Pack into an oiled loaf tin and cover with foil. Bake at 190°C/mark 5 for 30-40 minutes.

Weight loss: 23%

## 5 Beanburger, aduki, fried in vegetable oil

120g chopped onion
10g vegetable oil
320g boiled aduki beans
75g porridge oats
10g chopped fresh parsley

1tsp mixed herbs
20g soya sauce
3g tomato purée
1 egg
vegetable oil absorbed on frying (20g)

Fry onion in oil until brown. Mix beans and onions together with remaining ingredients. Form into 6-8 shapes approximately 1 cm thick. Fry for 3 minutes either side.

Weight loss: 9%


## 6 Beanburger, butter bean, fried in vegetable oil

270g canned butter beans
22g vegetable oil
90g finely chopped onion
90g grated carrot

75g porridge oats
1tsp marmite
2tsp mixed herbs
vegetable oil absorbed on frying (25g)

Mash or purée beans. Heat oil and fry onion and carrot for 3-5 minutes without browning. Mix beans, onions and carrots together with all other ingredients and shape into 6 burgers. Fry, turning once.

Weight loss: 11%


## 7 Beanburger, red kidney bean, fried in vegetable oil

270g canned red kidney beans
22g vegetable oil
90g finely chopped onion
90g grated carrot

75g porridge oats
1tsp marmite
2tsp mixed herbs
vegetable oil absorbed on frying (25g)

Method as beanburger, butter, fried in vegetable oil (No. 6).

Weight loss: 11%


## 8 Beanburger, soya, fried in vegetable oil

120g chopped onion
10g vegetable oil
320g boiled soya beans
75g porridge oats
10g chopped fresh parsley

1tsp mixed herbs
20g soya sauce
35g tomato purée
1 egg
vegetable oil absorbed on frying (20g)

Method as beanburger, aduki, fried in vegetable oil (No. 5).

Weight loss: 9%

## 9 Bhaji, aubergine and potato

360g peeled potatoes
200g aubergine
60g vegetable oil
½tsp cumin seeds
½tsp mustard seeds
220g chopped onion

16g tomato purée
½tsp salt
½tsp turmeric powder
½tsp chilli powder
1 tsp garam masala
90ml water

Chop the potatoes and aubergines into large chips. Heat oil and add cumin and mustard seeds. Add onions and cook until brown, add remaining ingredients except water and cook for a few minutes. Reduce the heat, add the water, cover and simmer for 30-35 minutes.

Weight loss: 25%

## 10 Bhaji, aubergine, pea, potato and cauliflower

11g vegetable oil
½tsp cumin seeds
260g diced aubergine
150g peas
150g diced potato
150g cauliflower florets

1tsp salt
1tsp chilli powder
1tsp turmeric powder
10ml lemon juice
1½tsp sugar
70ml water

Heat oil, add cumin seeds. When seeds begin to pop add vegetables and remaining ingredients. Cook until vegetables are tender.

Weight loss: 26%

## 11 Bhaji, cabbage

25g vegetable oil
5g dried red chillies
180g finely chopped onion

15g crushed garlic
540g shredded cabbage
3g salt

Heat oil and add chillies. Add onions and garlic and cook until brown. Add shredded cabbage and cook for 5 minutes stirring all the time. Stir in the salt and serve.

Weight loss: 25%

## 12 Bhaji, cabbage and pea, with butter ghee

410g chopped onions
50g chopped root ginger
115g butter ghee
450g shredded cabbage

225g frozen peas
1tsp salt
1tsp chilli powder
1tsp turmeric

Fry onions and ginger in butter ghee until brown. Add cabbage and frozen peas and stir for 5 minutes. Add salt, chilli powder and turmeric, cover and cook on a low heat for 15 minutes.

Weight loss: 35%

### 13 Bhaji, cabbage and pea, with vegetable oil

410g chopped onions
50g chopped root ginger
115g vegetable oil
450g shredded cabbage

225g frozen peas
1tsp salt
1tsp chilli powder
1tsp turmeric

Method as Bhaji, cabbage and pea, with butter ghee (No. 12).

Weight loss: 35%


### 14 Bhaji, cabbage and potato, with butter

1040g green cabbage
300g potato
155g tomato
260g onion

90g butter
15g root ginger
15g green chilli
20g garlic

Quantities are derived from dietary survey records.

Weight loss: 28%


### 15 Bhaji, cabbage and potato, with vegetable oil

1040g green cabbage
300g potato
155g tomato
260g onion

90g vegetable oil
15g root ginger
15g green chilli
20g garlic

Quantities are derived from dietary survey records.

Weight loss: 28%


### 16 Bhaji, cabbage and spinach

chopped cabbage
chopped spinach
water
chopped onion
vegetable oil
chick pea flour

crushed garlic
salt
red chilli powder
ginger
green chilli

Proportions obtained from a manufacturer.


### 17 Bhaji, carrot, potato and pea, with butter

780g carrots
340g potato
295g peas
260g onions

185g tomatoes
110g butter
12g garlic
20g root ginger

Quantities are derived from dietary survey records.

Weight loss: 27%

## 18  Bhaji, carrot, potato and pea, with vegetable oil

780g carrots  
340g potato  
295g peas  
260g onions  

185g tomatoes  
110g vegetable oil  
12g garlic  
20g root ginger  

Quantities are derived from dietary survey records.

Weight loss: 27%

## 19  Bhaji, cauliflower

100g chopped onions  
185g vegetable ghee  
1000g cauliflower florets  
10g chopped pepper  

2tsp chilli powder  
2tsp salt  
50ml water (adhering to washed cauliflower)  

Fry the onions in ghee until brown. Add remaining ingredients, cover and cook for 40 minutes.

Weight loss: 30%

## 20  Bhaji, cauliflower and potato

450g cauliflower florets  
30g vegetable oil  
4g cumin seeds  
180g diced boiled potatoes  
4g cumin powder  

2g coriander powder  
4g turmeric  
4g chilli powder  
3g salt  

Soak the cauliflower florets for 30 minutes and drain. Heat the oil and add whole cumin seeds. Add cauliflower and stir for about 2 minutes, let it brown in spots. Cover, reduce heat and simmer for 4-6 minutes or until the cauliflower has softened. Add remaining ingredients and continue to cook until potatoes are heated through.

Weight loss: 25%

## 21  Bhaji, cauliflower, potato and pea, with butter

cauliflower  
potatoes  
frozen peas  
salt  
green chillies  
root ginger  
tomatoes  

chilli powder  
turmeric  
cumin seeds  
honey  
butter  
lemon juice  

Proportions obtained from a manufacturer.

## 22 Bhaji, cauliflower, potato and pea, with vegetable oil

cauliflower  
potatoes  
frozen peas  
salt  
green chillies  
root ginger  
tomatoes  
chilli powder  
turmeric  
cumin seeds  
honey  
vegetable oil  
lemon juice  

Proportions obtained from a manufacturer.

## 23 Bhaji, cauliflower and vegetable

246g chopped onions  
126g butter  
946g cauliflower florets  
96g tomatoes  
36g chopped spring onions  
56g coriander leaves  
175ml water  
20g garam masala  
17g curry powder  
17g paprika  
14g salt  
20g crushed garlic  
56g sliced green chillies  

Fry onions in butter until brown. Add remaining ingredients, cover and cook for 40 minutes.

Weight loss: 30%

## 24 Bhaji, green bean

green beans  
salt  
crushed garlic  
crushed ginger  
tomato purée  
tomatoes  
chilli powder  
turmeric  
mustard seeds  
fenugreek seeds  
cumin seeds  
coriander seeds  
honey  
vegetable oil  
lemon juice  

Proportions obtained from a manufacturer.

## 25 Bhaji, karela, with butter ghee

87% boiled karela  
10% butter ghee  
3% mixed spices  

Proportions are derived from dietary survey records.

## 26 Bhaji, karela, with vegetable oil

87% boiled karela  
10% vegetable oil  
3% mixed spices  

Proportions are derived from dietary survey records.

## 27 Bhaji, mushroom

300g finely sliced onions
100g vegetable oil
1tsp chilli powder

2g crushed garlic
1tsp turmeric
350g mushrooms

Fry onions in oil until soft. Add chilli powder, garlic and turmeric and cook for 5 minutes. Add mushrooms and continue until mushrooms are cooked.

Weight loss: 19%

## 28 Bhaji, mustard leaves

450g finely chopped mustard leaves
7g salt
300ml water
5g chopped root ginger
5g chopped green chillies

30g chopped onion
30g butter
2g crushed garlic
4g mixed spices

Cook mustard leaves, salt and water in a pressure cooker at 15lb pressure for 20 minutes. Add ginger, chillies and onion to spinach and boil off water. Pound mixture to a soft consistency and fry in butter with garlic and spices until all the water has evaporated.

Weight loss: 49%

## 29 Bhaji, mustard leaves and spinach

285g finely chopped mustard leaves
165g finely chopped spinach
7g salt
300ml water
5g chopped root ginger

5g chopped green chillies
30g chopped onion
30g butter
2g crushed garlic
4g mixed spices

Method as bhaji, mustard leaves (No.28).

Weight loss: 49%

## 30 Bhaji, okra, Bangladeshi, with butter ghee

450g finely chopped onion
5g crushed garlic
35g butter ghee
340g trimmed okra

10g chopped green chillies
½tsp salt
½tsp turmeric

Fry onion and garlic in the ghee until brown. Cut okra into ½ inch pieces and add to onion mix. Sprinkle with chillies, salt and turmeric. Simmer with stirring for 10 minutes.

Weight loss: 27%

## 31 Bhaji, okra, Bangladeshi, with vegetable oil

450g finely chopped onion
5g crushed garlic
35g vegetable oil
340g trimmed okra

10g chopped green chillies
½tsp salt
½tsp turmeric

Method as bhaji, okra, Bangladeshi, with butter ghee (No. 30).

Weight loss: 27%

## 32 Bhaji, okra, Islami

| | |
|---|---|
| 160g chopped onions | ½tsp chilli powder |
| 44g vegetable oil | 12g tomato purée |
| 370g trimmed okra | 1tsp coriander powder |
| ¾tsp salt | 1tsp cumin powder |
| ½tsp garlic paste | ¼tsp turmeric |

Fry onion in oil. Slice okra lengthwise and add to onions, cook in a low heat for 30 minutes. Add remaining ingredients and stir. The curry is ready when oil starts to rise to the surface and the texture is dry.

Weight loss: 50%

## 33 Bhaji, pea

| | |
|---|---|
| 100g chopped onions | 150g butter |
| 2g turmeric | 7g salt |
| 2g garam masala | 225g fresh peas |
| 2g chilli powder | 10g chopped coriander leaves |

Fry onion and spices in the butter until onion is soft. Add salt, peas and coriander leaves. Cover and simmer until tender.

Weight loss: 25%

## 34 Bhaji, potato, with butter ghee

| | |
|---|---|
| 82g chopped onion | 4g turmeric |
| 11g chopped coriander leaves | 10g chopped green chillies |
| 50g butter ghee | 50ml water |
| 454g diced potatoes | 6g garam masala |
| 5g salt | 4ml lemon juice |

Fry onion and coriander in the ghee. Add potatoes, salt, turmeric, chillies and water. Cover and cook for 30 minutes. Add garam masala and lemon juice 10 minutes before end of cooking.

Weight loss: 23%

## 35 Bhaji, potato, with vegetable oil

| | |
|---|---|
| 82g chopped onion | 4g turmeric |
| 11g chopped coriander leaves | 10g chopped green chillies |
| 50g vegetable oil | 50ml water |
| 454g diced potatoes | 6g garam masala |
| 5g salt | 4ml lemon juice |

Method as bhaji, potato, with butter ghee (No. 34).

Weight loss: 23%

## 36 Bhaji, potato and fenugreek leaves

44g oil
2tsp cumin seeds
240g peeled and diced potatoes
80g chopped fenugreek leaves
2tsp chilli powder

1tsp ground coriander
1tsp cumin powder
2tsp sugar
20ml lemon juice

Heat oil and add cumin seeds. Add potato and cook for 5 minutes. Add fenugreek leaves, chilli, coriander, cumin and sugar. Mix well, reduce heat, cover and cook for a further 6-8 minutes. Add lemon juice and serve.

Weight loss: 27%

## 37 Bhaji, potato and green pepper

415g green pepper
650g potato
190g onion
110g vegetable oil

165g tomatoes
30g chilli
5g root ginger

Quantities are derived from dietary survey records.

Weight loss: 20%

## 38 Bhaji, potato and onion

225g chopped onions
2g cumin seeds
55g butter

340g boiled potatoes
½tsp salt
10g chopped green chillies

Fry the onion and cumin seeds in butter until lightly brown. Cut potatoes into fairly large pieces and add to the onions together with the salt and chillies. Simmer for a few minutes until well mixed.

Weight loss: 28%

## 39 Bhaji, potato, onion and mushroom

330g finely sliced onions
250g finely sliced potatoes
100g vegetable oil
1tsp chilli powder

5g crushed garlic
1tsp turmeric
100g mushrooms

Fry onions and potatoes until soft. Add chilli powder, garlic and turmeric and cook for 5 minutes. Add mushrooms and continue until mushrooms are cooked.

Weight loss: 26%

## 40 Bhaji, potato, spinach and cauliflower

30.3% boiled potatoes
22.8% boiled spinach
22.8% boiled cauliflower
13.6% butter ghee

7.6% fried onions
1.5% garlic
0.8% garam masala
0.6% chilli powder

Proportions are derived from dietary survey records.

## 41  Bhaji, spinach

| | |
|---|---|
| 450g finely chopped spinach | 30g chopped onion |
| 7g salt | 30g butter |
| 300ml water | 2g crushed garlic |
| 5g chopped root ginger | 4g mixed spices |
| 5g chopped green chillies | |

Method as bhaji, mustard leaves (No. 28).

Weight loss: 49%

## 42  Bhaji, spinach and potato

| | |
|---|---|
| 225g potatoes, cut into quarters | 10g green chillies |
| 44g vegetable oil | ½tsp chilli powder |
| 10g crushed root ginger | 225g chopped spinach |
| 8g crushed garlic | ½tsp salt |

Fry potatoes in oil until brown. In a separate pan fry ginger, garlic, chillies and chilli powder. Add spinach and fry, add potatoes and salt. Cover and simmer until potatoes are cooked.

Weight loss: 37%

## 43  Bhaji, turnip

| | |
|---|---|
| 42g chopped onion | 2g turmeric |
| 5g chopped coriander leaves | 4ml lemon juice |
| 4g ground ginger | 370g diced turnip |
| 60g butter ghee | 2g garam masala |
| 5g salt | |

Fry onion, coriander leaves and ginger in ghee. Add salt, turmeric and lemon juice. Let this dry up and then add turnip. Cook, stirring for 5 minutes, cover and cook for 30-40 minutes on a low heat. Add garam masala 10 minutes before the end of cooking.

Weight loss: 25%

## 44  Bhaji, turnip and onion

| | |
|---|---|
| 690g turnip | 10g garlic |
| 570g onion | 10g root ginger |
| 120g ghee | 15g green chillies |

Quantities are derived from dietary survey records.

Weight loss: 20%

## 45 Bhaji, vegetable, with butter

150g chopped onion
200g butter
5g ground ginger
2g crushed garlic
5g chillies
4g mixed spices
7g salt

300g diced potatoes
300g sliced carrots
120g diced turnip
200g sliced aubergine
200g cauliflower florets
90g tomatoes
30g peas

Fry onion in butter until brown. Add ginger, garlic and remaining spices and stir. Add vegetables and simmer for 10 minutes or until tender.

Weight loss: 31%

## 46 Bhaji, vegetable, with vegetable oil

150g chopped onion
200g vegetable oil
5g ground ginger
2g crushed garlic
5g chillies
4g mixed spices
7g salt

300g diced potatoes
300g sliced carrots
120g diced turnip
200g sliced aubergine
200g cauliflower florets
90g tomatoes
30g peas

Method as bhaji, vegetable, with butter (No. 45).

Weight loss: 31%

## 47 Bhaji, vegetable, Punjabi, with butter

onion
cauliflower
peas
carrots
green beans
tomatoes
butter
tomato purée

salt
garlic
cumin
turmeric
green chillies
root ginger
chilli powder
ground ginger

Proportions obtained from a manufacturer.

## 48 Bhaji, vegetable, Punjabi, with vegetable oil

onion
cauliflower
peas
carrots
green beans
tomatoes
vegetable oil
tomato purée

salt
garlic
cumin
turmeric
green chillies
root ginger
chilli powder
ground ginger

Proportions obtained from a manufacturer.

### 49 Broccoli in cheese sauce, made with whole milk

560g broccoli, cut into florets
400ml cheese sauce made with whole milk
40g grated cheese

Boil the broccoli until just tender. Drain and cover with cheese sauce. Sprinkle with grated cheese and brown under a grill or in a hot oven, 220°C/mark 7.

Weight loss: 8%

### 50 Broccoli in cheese sauce, made with semi-skimmed milk

560g broccoli, cut into florets
400ml cheese sauce made with semi-skimmed milk
40g grated cheese

Method as broccoli in cheese sauce, made with whole milk (No. 49).

Weight loss: 8%

### 51 Broccoli in cheese sauce, made with skimmed milk

560g broccoli, cut into florets
400ml cheese sauce made with skimmed milk
40g grated cheese

Method as broccoli in cheese sauce, made with whole milk (No. 49).

Weight loss: 8%

### 52 Bubble and squeak, fried in lard

46% boiled cabbage
46% boiled potato
lard absorbed on frying (8%)

Fry the cabbage and potato together.

Weight loss: 10%

### 53 Bubble and squeak, fried in sunflower oil

46% boiled cabbage
46% boiled potato
sunflower oil absorbed on frying (8%)

Fry cabbage and potato together.

Weight loss: 10%

### 54 Bubble and squeak, fried in vegetable oil

46% boiled cabbage
46% boiled potato
vegetable oil absorbed on frying (8%)

Fry the cabbage and potato together.

Weight loss: 10%

## 55  Cabbage, red, cooked with apple

900g shredded red cabbage    40g sugar
40g margarine    70ml water
80g finely chopped onion    70ml vinegar
200g peeled, cored and chopped baking apples

Fry the cabbage in the margarine. Add the remaining ingredients and cover. Simmer gently for 35 minutes, stirring occasionally.

Weight loss: 10%


## 56  Callaloo and cho cho

85g diced tomato    5g black pepper
80g chopped onion    5g mixed spices
15g vegetable oil    455g spinach(callaloo)
100g peeled and chopped cho cho    150ml water
5g salt    40g butter

Fry tomato and onion in the oil for 4 minutes. Add cho cho, salt, pepper and spices. Add spinach (callaloo), water and butter. Cover and cook for 15 minutes.

Weight loss: 15%


## 57  Callaloo and okra

60g chopped tomato    60g trimmed okra
45g chopped onion    225g spinach
2g crushed garlic    5g salt
10g vegetable oil    150ml water
25g butter

Fry tomato, onion and garlic in oil and butter for 4 minutes. Add okra and spinach, salt and water. Cover and cook for 15-20 minutes.

Weight loss: 11%


## 58  Cannelloni, spinach

31% milk    3% flour
26% boiled pasta    2% breadcrumbs
16% boiled spinach    2% Parmesan
10% Ricotta cheese    1.5% cornflour
4% boiled onions    0.5% salt
4% vegetable oil

Proportions are derived from dietary survey records.

### 59 Cannelloni, vegetable

30% milk
25% boiled pasta
10% Ricotta cheese
5% tomatoes
4% boiled onions
4% vegetable oil
3% flour
2% boiled carrots
2% boiled spinach

2% boiled courgettes
2% boiled cabbage
2% boiled leeks
2% breadcrumbs
2% butter
2% Parmesan cheese
1.5% cornflour
0.5% salt

Proportions are derived from dietary survey records.

### 60 Casserole, bean and mixed vegetable

250g beans
100g chopped onion
60g chopped celery
225g sliced carrot
120g sliced red pepper

120g sliced green pepper
120g canned sweetcorn
30g tomato purée
600ml water
2tsp marmite

Place all ingredients in a casserole and stir. Cover and cook for approximately 1 hour at 190°C/mark 5.

Weight loss: 15%

### 61 Casserole, bean and root vegetable

250g beans
150g diced potatoes
100g diced carrots
100g diced swede/turnip

100g chopped onion
400g canned tomatoes
300ml water
1tsp marmite

Method as casserole, bean and mixed vegetable (No. 60).

Weight loss: 15%

### 62 Casserole, sweet potato and green banana

780g boiled sweet potatoes
112g butter
10g garlic salt
1½tsp pepper

430g peeled green bananas
170g sugar
110ml freshly squeezed orange juice

Thinly slice sweet potatoes and line a buttered casserole dish. Sprinkle with salt and pepper, dot with butter and add a layer of thinly sliced bananas. Sprinkle with sugar and continue adding layers of potato and banana, finishing with banana. Sprinkle with sugar, dot with butter, pour over orange juice and bake at 180°C/mark 4 for 30 minutes.

Weight loss: 10%

### 63 Casserole, vegetable

240g diced potato
120g sliced carrot
120g diced onion
120g diced swede
120g diced parsnip

90g canned sweetcorn
90g frozen peas
90g chopped tomatoes
450g canned tomatoes
1tsp marmite

Method as casserole, bean and mixed vegetable (No. 60).

Weight loss: 15%

### 64 Cauliflower cheese, made with whole milk

100g grated cheese
1 small cauliflower (700g)
100ml cauliflower water
½ level tsp salt

25g margarine
25g flour
250ml whole milk
pepper

Boil cauliflower until just tender, break into florets. Drain saving 100ml water, place in a dish and keep warm. Make a white sauce from the margarine, flour milk and cauliflower water. Add 75g cheese and season. Pour over the cauliflower and sprinkle with the remaining cheese. Brown under a grill or in a hot oven, 220°C/mark 7.

Weight loss: 15%

### 65 Cauliflower cheese, made with semi-skimmed milk

100g grated cheese
1 small cauliflower (700g)
100ml cauliflower water
½ level tsp salt

25g margarine
25g flour
250ml semi-skimmed milk
pepper

Method as cauliflower cheese, made with whole milk (No. 64).

Weight loss: 15%

### 66 Cauliflower cheese, made with skimmed milk

100g grated cheese
1 small cauliflower (700g)
100ml cauliflower water
½ level tsp salt

25g margarine
25g flour
250ml skimmed milk
pepper

Method as cauliflower cheese, made with whole milk (No. 64).

Weight loss: 15%

### 67 Cauliflower in white sauce, made with whole milk

67% boiled cauliflower

33% white sauce made with whole milk

Proportions are derived from dietary survey records.

## 68 Cauliflower in white sauce, made with semi-skimmed milk

67% boiled cauliflower      33% white sauce made with semi-skimmed milk

Proportions are derived from dietary survey records.


## 69 Cauliflower in white sauce, made with skimmed milk

67% boiled cauliflower      33% white sauce made with skimmed milk

Proportions are derived from dietary survey records.


## 70 Cauliflower with onions and chilli pepper

22g vegetable oil      15g chopped green chillies
180g chopped onions      700g boiled cauliflower (1 small)

Fry onions in the oil until brown, add chilli and remove from heat. Cover the warm cauliflower with fried onions and chilli and serve.

Weight loss: 8%


## 71 Chilli, bean and lentil

50g  chopped onion      200g boiled or canned red kidney beans
50g red chopped pepper      100g boiled whole lentils
10g vegetable oil      5g paprika
5g garlic      5g chilli powder
300g canned tomatoes      2g salt

Fry onion and pepper in oil until soft. Add remaining ingredients and simmer for a minimum of 20 minutes.

Weight loss: 25%


## 72 Chilli, Quorn

200g Quorn      20g tomato purée
100g chopped onion      200g canned tomatoes
80g chopped red pepper      1tsp sugar
10g crushed garlic      1tsp wine vinegar
22g vegetable oil      250ml water
1tsp chilli powder      2tsp cornflour
½tsp nutmeg      75g canned red kidney beans

Gently fry the Quorn, onion, pepper and garlic in oil for 3-4 minutes. Add chilli powder and cook for 1 minute. Add nutmeg, tomatoes and tomato purée, sugar, vinegar and water and simmer for 10 minutes. Mix cornflour to a paste, add to sauce and stir. Add kidney beans and cook for a further 3 minutes.

Weight loss: 24%

## 73 Chilli, vegetable

| | |
|---|---|
| 120g onion | 440g boiled or canned red kidney beans |
| 240g carrots | 10g chilli powder |
| 240g parsnips | 330g canned sweetcorn |
| 120g pepper | 14g oxo |
| 180g courgette | 568ml water |
| 400g canned tomatoes | 5g salt |

Quantities are derived from dietary survey records.

Weight loss: 15%

## 75 Cho cho fritters, fried in vegetable oil

| | |
|---|---|
| 190g grated cho cho flesh | 5g butter |
| 50g flour | 1 beaten egg |
| 55ml milk | vegetable oil absorbed on frying (27g) |
| ½tsp salt | |

Mix together the cho cho, flour, milk, salt and melted butter. Fold beaten egg into the mixture. Drop tablespoonfuls into hot fat and fry until brown. Drain.

Weight loss: 20%

## 76 Coco fritters, fried in vegetable oil

| | |
|---|---|
| 310g grated coco | ½tsp pepper |
| 50g flour | 2 beaten eggs |
| 2tsp baking powder | vegetable oil absorbed on frying (95g) |
| 1tsp salt | |

Mix coco together with flour, baking powder, salt and pepper. Fold beaten eggs into mixture. Drop tablespoonsful into hot fat and fry until brown.

Weight loss: 17%

## 77 Coleslaw, with mayonnaise, retail

| | |
|---|---|
| cabbage | carrot |
| mayonnaise | onion |

Proportions obtained from a manufacturer

## 78 Coleslaw, with reduced calorie dressing, retail

| | |
|---|---|
| cabbage | carrot |
| reduced calorie dressing | onion |

Proportions obtained from a manufacturer

## 79 Coleslaw with vinaigrette, retail

| | |
|---|---|
| cabbage | carrots |
| vinaigrette dressing | onions |

Proportions obtained from a manufacturer

## 80 Coo-coo

60g okra
1000ml water
7g salt

300g cornmeal
15g butter

Wash and slice okra, add salt and boil in half the water for 10 minutes. Mix cornmeal with remaining water and blend, along with butter, into okra mix. Continue to stir until mixture becomes stiff and breaks from bottom of the pan.

Weight loss: 9%


## 81 Corn fritters, fried in vegetable oil

2 beaten eggs
142ml milk
225g flour
7g salt
½tsp pepper

2tsp baking powder
310g canned sweetcorn
15g melted butter
vegetable oil absorbed on frying (90g)

Mix eggs with milk. Gradually add remaining ingredients and mix, saving baking powder until last. Drop tablespoonsful into hot oil and fry until brown. Drain.

Weight loss: 10%


## 82 Corn pudding

450g canned sweetcorn
437ml milk
3 beaten eggs

70g chopped onion
½tsp salt

Combine all ingredients and pour into a lightly oiled casserole. Place casserole in a shallow pan filled with hot water. Bake at 180°C/mark 4 for 45 minutes or until a knife can be pulled cleanly from centre.

Weight loss: 11%


## 83 Courgettes with eggs

500g diced courgettes
45g chopped onion
44g corn oil

2 well beaten eggs
¼tsp salt

Fry the courgettes and onion in corn oil. Add eggs and seasoning, stir gently until courgettes and eggs blend and the egg is cooked.

Weight loss: 18%

## 84  Crumble, vegetable in milk base

Base:
360g cauliflower
22g vegetable oil
80g flour
350ml milk
150ml water
326g canned sweetcorn
10g chopped fresh parsley
60g grated cheese

Topping:
150g flour
65g margarine
2g salt

Cook the cauliflower for 5 minutes saving the water after draining. Prepare the sauce with oil, flour, milk and vegetable water. Stir in sweetcorn, parsley and cheese. Fold in cauliflower and turn into a shallow dish. Rub flour and salt with margarine and sprinkle over vegetable mixture. Bake at 190°C/mark 5 for 30 minutes.

Weight loss: 15%

## 85  Crumble, vegetable in milk base, wholemeal

Base:
360g cauliflower
22g vegetable oil
80g wholemeal flour
350ml milk
150ml water
326g canned sweetcorn
10g chopped fresh parsley
60g grated cheese

Topping:
150g wholemeal flour
65g margarine
2g salt

Method as crumble, vegetable in milk base (No. 84).

Weight loss: 15%

## 86  Crumble, vegetable in tomato base

Base:
90g chopped onion
90g sliced carrots
22g vegetable oil
90g cauliflower florets
90g sliced courgettes
425g canned tomatoes

Topping:
125g flour
55g margarine
2g salt

Fry the onion and carrots. Add remaining vegetables and canned tomatoes, season and cook for 15 minutes or until the vegetables are tender. Pour into a shallow casserole dish. Rub flour and salt with margarine. Sprinkle over vegetable mixture and bake at 200°C/mark 6 for 30 minutes.

Weight loss: 31%

## 87 Crumble, vegetable in tomato base, wholemeal

Base:
90g chopped onion
90g sliced carrots
22g vegetable oil
90g cauliflower florets
90g sliced courgettes
425g canned tomatoes

Topping:
125g wholemeal flour
55g margarine
2g salt

Method as crumble, vegetable in tomato base (No. 86).

Weight loss: 31%

## 88 Curry, almond

60g crushed garlic
42g chopped green chillies
150g butter ghee
120g chopped almonds
30g crushed root ginger

24g cumin seeds
6g curry powder
4g salt
6g chilli powder
1000ml water

Fry garlic and chillies in the ghee. Add almonds and remaining ingredients. Cook for 1-1½ hours.

Weight loss: 54%

## 89 Curry, aubergine

200g chopped onions
1tsp salt
58g vegetable oil
5g crushed garlic
10g crushed root ginger

220g canned tomatoes
1tsp chilli powder
200g sliced aubergine
50g sliced potato
50ml water

Fry the onions and salt until brown. Add garlic, ginger, tomatoes and chilli powder and cook until well blended and the oil starts to rise to the surface. Add aubergine, potato and water, cover and simmer until vegetables are cooked.

Weight loss: 26%

## 90 Curry, black gram dahl

240g dry black gram dahl
215ml water absorbed on soaking
1680ml water added during cooking
60g butter
68g chopped onion
5g root ginger

2g garlic
4g chilli powder
2g coriander powder
2g cumin powder
7g salt

Soak dahl in water for 1 hour. Drain, add remaining ingredients and either simmer over a low heat until tender or cook in a pressure cooker for 10 minutes at 15lb pressure.

Weight loss: 30%

## 91 Curry, black gram, whole, Bengali

220g dry whole black gram
120ml water, absorbed on soaking
1000ml water added during cooking
120g chopped onion
100g chopped tomatoes
40g butter

5g crushed root ginger
2g crushed garlic
4g mixed spices
7g salt
2tsp sugar
10ml lemon juice

Soak the beans for 12 hours. Drain, add remaining ingredients and either simmer over a low heat until tender or cook in a pressure cooker for 20 minutes at 15lb pressure.

Weight loss: 19%

## 92 Curry, black gram, whole, Gujerati

250g dry whole black gram
520ml water absorbed on
  soaking and boiling
150g chopped onion
60g butter
7g chopped green chilli

7g crushed root ginger
5g crushed garlic
2tsp salt
1tsp turmeric
55g tomato purée
1tsp garam masala

Wash, soak and boil black gram until soft. Fry onion in the butter, add green chilli, ginger and garlic. Add salt, turmeric and tomato purée and cook until the fat rises to the surface. Add the drained boiled beans to the sauce and mix well. Add garam masala and heat.

Weight loss: 24%

## 93 Curry, black gram, whole, and red kidney bean

85g dry whole black gram
25g dry red kidney beans
1800ml water
90g chopped onion
14g chopped green chillies
1tsp salt

30g vegetable oil
½tsp cumin seeds
2tsp tomato purée
½tsp turmeric
1tsp chilli powder
1tsp garam masala

Wash the beans, add water, onion, chillies and salt and bring to the boil. Reduce heat and boil gently for approximately 2 hours. In a separate pan heat oil, add remaining ingredients and simmer. Add to bean mixture and simmer for a further 3-5 minutes.

Weight loss: 68%

## 94 Curry, black-eye bean, Gujerati

225g dry small black-eye
  beans (chora)
315ml water absorbed on
  boiling beans
1tsp salt
25g vegetable oil

6g crushed garlic
1tsp turmeric
1tsp chilli powder
240ml water added as stock
10ml lemon juice

Soak beans and boil until soft in salted water. Heat oil in pan, add garlic and remaining spices. Add beans and stock, when all water has evaporated add lemon juice and stir.

Weight loss: 19%


## 95 Curry, black-eye bean, Punjabi

66.1% boiled black-eye beans
16.4% water
5.7% boiled onion
5.7% tomato

4.3% butter
1.0% crushed root ginger
0.3% crushed garlic
0.3% crushed chilli

Proportions are derived from dietary survey records.


## 96 Curry, Bombay potato

30g vegetable oil
1tsp cumin seeds
½tsp mustard seeds
360g diced potatoes
½tsp turmeric

1tsp salt
½tsp chilli powder
115g finely chopped tomatoes
150ml water

Heat the oil and add cumin and mustard seeds. Leave for a few seconds then add potatoes, turmeric, salt, chilli powder and tomatoes. Cook for 5 minutes then add water, bring to the boil, lower heat and simmer gently for 20 minutes.

Weight loss: 28%


## 97 Curry, cabbage

200g chopped onions
1tsp salt
32g vegetable oil
5g crushed garlic
10g crushed root ginger

220g canned tomatoes
1tsp chilli powder
50g sliced potato
450g chopped white cabbage

Fry the onions and salt in oil until brown. Add garlic, ginger, tomatoes and chilli powder and cook until well blended and the oil starts to rise to the surface. Add potato. When potato is cooked add cabbage and simmer until soft.

Weight loss: 30%

## 98 Curry, cauliflower and potato

100g sliced onions
30g vegetable oil
20g crushed root ginger
2tsp salt
1tsp chilli powder
1tsp turmeric
400g peeled potatoes

270ml water
1500g cauliflower florets
100g canned tomatoes
20g tomato purée
1tsp garam masala
5g chopped coriander leaves

Fry onions until brown, add ginger, salt, chilli powder and turmeric. Mix well. Cut potatoes into medium sized pieces and add to onion mixture together with water. Cook for 5 minutes. Cut cauliflower into medium sized pieces and add to mixture. Purée tinned tomatoes and mix with tomato purée. When vegetables are cooked add tomato mixture and garam masala, mix well and simmer for 2 minutes. Add coriander leaves.

Weight loss: 21%

## 99 Curry, chick pea dahl

225g dry chick pea dahl
200ml water absorbed on soaking
28g vegetable oil
60g chopped onion
2g crushed garlic

1tsp chilli powder
½tsp garam masala
7g chopped green chilli
100g chopped tomato
415ml water

Soak the chick pea dahl overnight. Fry the onion and garlic in the oil until brown. Add a little water together with spices and tomatoes. Stir and cook until dry. Add dahl and water, simmer until cooked.

Weight loss: 35%

## 100 Curry, chick pea dahl and spinach, with butter

36.8% water
24.5% dry chick pea dahl
24.5% boiled spinach

12.3% butter
0.8% curry powder
1.1% green chillies

Proportions are derived from dietary survey records.

Weight loss: 2%

## 101 Curry, chick pea dahl and spinach, with vegetable oil

36.8% water
24.5% dry chick pea dahl
24.5% boiled spinach

12.3% vegetable oil
0.8% curry powder
1.1% green chillies

Proportions are derived from dietary survey records.

Weight loss: 2%

102 **Curry, chick pea, whole**

225g dry small whole chick peas
235ml water absorbed on boiling
25g vegetable oil
10g cumin seeds
6g crushed garlic

1tsp salt
1tsp turmeric
1tsp chilli powder
215ml water added as stock
20ml lemon juice

Soak chick peas and boil until soft then drain. Heat oil, add garlic and cumin seeds and fry until brown. Add remaining ingredients except lemon juice, stir well and cook. Add the lemon juice and stir.

Weight loss: 29%


103 **Curry, chick pea, whole, basic**

180g chopped onion
20g vegetable oil
½tsp chilli powder
1tsp ground cumin
½tsp ground coriander

½tsp ground ginger
250g chopped tomatoes
20g tomato purée
650g boiled whole chick peas
450ml water

Fry onions in the oil until soft. Add spices and cook for a further 2 minutes. Add remaining ingredients. Cover and simmer for 1 hour.

Weight loss: 36%


104 **Curry, chick pea, whole and potato**

30g vegetable oil
30g chick pea flour
150g natural yogurt
300ml water
360g boiled small whole chick peas

50g diced boiled potatoes
½tsp turmeric
7g chopped green chilli
7g chopped root ginger
5g crushed garlic

Heat oil, add chick pea flour and roast until light brown. Add yogurt and water, mix well. Add remaining ingredients and simmer for 15 minutes.

Weight loss: 14%


105 **Curry, chick pea, whole and tomato, Gujerati, with butter ghee**

225g dry whole chick peas
265ml water
85g butter ghee
90g minced onion
5g crushed garlic
7g crushed root ginger
7g chopped green chilli

2tsp salt
½tsp turmeric
1tsp garam masala
250g thinly sliced tomatoes
1½tsp sugar
10ml lemon juice

Soak chick peas overnight. Boil until soft. Fry onion, garlic, ginger, chilli and seasoning until onions are soft. Add tomatoes and sugar and cook. Add chick peas together with water and simmer for at least 30 minutes. Add lemon juice, mix and serve.

Weight loss: 30%

### 106 Curry, chick pea, whole and tomato, Gujerati, with vegetable oil

225g dry whole chick peas
265ml water
85g vegetable oil
90g minced onion
5g crushed garlic
7g crushed root ginger
7g chopped green chilli

2tsp salt
½tsp turmeric
1 tsp garam masala
250g thinly sliced tomatoes
1½tsp sugar
10ml lemon juice

Method as curry, chick pea, whole and tomato, Gujerati, with butter ghee (No. 105).

Weight loss: 30%


### 107 Curry, chick pea, whole and tomato, Punjabi, with butter

64.5% boiled whole chick peas
15.0% water
8.0% onion
7.7% tomato

3.5% butter
0.6% root ginger
0.4% green chilli
0.3% garlic

Proportions are derived from dietary survey records.


### 108 Curry, chick pea, whole and tomato, Punjabi, with vegetable oil

64.5% boiled whole chick peas
15.0% water
8.0% onion
7.7% tomato

3.5% vegetable oil
0.6% root ginger
0.4% green chilli
0.3% garlic

Proportions are derived from dietary survey records.


### 109 Curry, courgette and potato

545g courgette
465g potato
65g vegetable oil

15g garlic
250g tomatoes
7g fresh chilli

Quantities are derived from dietary survey records.

Weight loss: 14%


### 110 Curry, dudhi, kofta

575g grated dudhi(bottle gourd) flesh
5g crushed green chilli
5g crushed root ginger
3tsp salt
1tsp turmeric
100g chick pea flour

30g butter ghee
450g chopped onion
300g canned tomatoes
284ml water
vegetable oil absorbed on frying
 dudhi balls (50g)

Squeeze grated gourd to remove water. Add chilli, ginger, 2tsp salt, ½tsp turmeric and chick pea flour. Mix well to form a smooth dough. Make walnut size balls and shallow fry in oil until light brown, drain and cool. Heat ghee, fry onion, add tomatoes and cook until the ghee rises to the top. Add remaining salt and turmeric together with water and simmer for 5 to 10 minutes. Add the fried gourd balls and mix well.

Weight loss: 33% for frying balls

38% for curry base

### 111　Curry, egg, with butter

| | |
|---|---|
| 115g butter | 1tsp chilli powder |
| 5g crushed garlic | 2tsp mixed spice |
| 50g chopped root ginger | 1tsp turmeric |
| 610g chopped onions | 425ml water |
| 235g canned tomatoes | 8 hard boiled eggs |
| 1tsp salt | |

Heat butter and fry garlic until brown. Add ginger and onions and fry until dark brown. Mash tomatoes into mixture. Add salt, chilli powder, spices, turmeric and mix well. Add water and boil for 15 minutes. Add eggs, simmer for a few minutes.

Weight loss: 31%

### 112　Curry, egg, with vegetable oil

| | |
|---|---|
| 115g butter | 1tsp chilli powder |
| 5g crushed garlic | 2tsp mixed spice |
| 50g chopped root ginger | 1tsp turmeric |
| 610g chopped onions | 425ml water |
| 235g canned tomatoes | 8 hard boiled eggs |
| 1tsp salt | |

Method as curry, egg, with butter (No. 111).

Weight loss: 31%

### 113　Curry, egg in sweet sauce

| | |
|---|---|
| 4 hard-boiled eggs | 300g sweet curry sauce |

Proportions are derived from a recipe review.

### 114　Curry, egg and potato

| | |
|---|---|
| 90g finely chopped onion | ½tsp chilli powder |
| 30g vegetable oil | 1tsp cumin powder |
| 110g peeled tomatoes | 180g diced potatoes |
| 1tsp salt | 6 hard boiled eggs |
| 1tsp turmeric | 150ml water |
| 1tsp garam masala | |

Fry the onion until brown. Add tomatoes and cook for 2 minutes. Add salt and spices and cook for a further 1 minute. Add potatoes and cook for 2 minutes followed by eggs cooking for a further 3 to 4. Add water, bring to the boil, lower heat and let simmer for 20 minutes.

Weight loss: 25%

## 115  Curry, gobi aloo sag, retail

cauliflower  
onion  
potato  
tomato  
spinach  
vegetable oil  
water  
salt  
sugar  
coriander  
garlic purée

Proportions obtained from a manufacturer.

Weight loss: 7% on re-heating

## 116  Curry, green bean

60g vegetable oil  
½tsp black mustard seeds  
½tsp cumin seeds  
½tsp asafoetida  
55g finely chopped tomatoes  
½tsp salt  
½tsp turmeric  
½tsp cumin powder  
½tsp coriander powder  
½tsp chilli powder  
340g thawed frozen green beans  
120ml water  
1tsp sugar  
5ml lemon juice

Heat oil and add mustard seeds, cumin seeds and asafoetida and cook for a few seconds. Add tomatoes, salt and remaining spices. Cook for a few minutes. Add beans, mix well and add water. Bring to the boil, lower heat and let simmer for 8 minutes. Add sugar and lemon juice.

Weight loss: 16%

## 117  Curry, karela

480g karela  
305g onion  
180g tomatoes  
70g corn oil

Quantities are derived from dietary survey records.

Weight loss: 29%

## 118  Curry, lentil, red/masoor dahl, with butter

225g dry red lentils/masoor dahl  
600ml water  
1tsp salt  
1tsp turmeric  
½tsp chilli powder  
55g butter  
180g chopped onion  
2g crushed garlic  
28g crushed root ginger

Place lentils with water, salt, turmeric and chilli powder in a pan. Bring to the boil and simmer for 30 minutes. Heat butter in a separate pan and fry onion, garlic and ginger until brown. Add to the lentil mixture and serve.

Weight loss: 32%

### 119 Curry, lentil, red/masoor dahl, with vegetable oil

225g dry red lentils/masoor dahl
600ml water
1tsp salt
1tsp turmeric
½tsp chilli powder

55g vegetable oil
180g chopped onion
2g crushed garlic
28g crushed root ginger

Method as curry, lentil, red/masoor dahl, with butter (No. 118).

Weight loss: 32%


### 120 Curry, lentil, red/masoor dahl, Punjabi

140g chopped onion
75g dry red lentils/masoor dahl
310ml water

1tsp salt
11g vegetable oil
1tsp chilli powder

Place lentils with water, salt and half the quantity of onion, in a pan. Bring to the boil and simmer for 30 minutes. Heat oil in a separate pan and fry remaining onion and chilli powder. Add to cooked lentils and heat through.

Weight loss: 49%


### 121 Curry, lentil, red/masoor dahl and tomato, with butter

170g dry red lentils/masoor dahl
1200ml water
90g chopped onion
1tsp salt
60g butter

½tsp cumin seeds
55g peeled tomatoes
1tsp turmeric
½tsp garam masala
½tsp chilli powder

Place lentils with water, onion and salt. Bring to the boil and simmer for 30 minutes. Heat butter in a separate pan, add remaining ingredients and cook for 2-3 minutes. Add the lentil mixture to the spices and simmer for 5 minutes.

Weight loss: 37%


### 122 Curry, lentil, red/masoor dahl and tomato, with vegetable oil

170g dry red lentils/masoor dahl
1200ml water
90g chopped onion
1tsp salt
60g vegetable oil

½tsp cumin seeds
55g peeled tomatoes
1tsp turmeric
½tsp garam masala
½tsp chilli powder

Method as curry, lentil red/masoor dahl and tomato, with butter (No. 121).

Weight loss: 37%

### 123 Curry, lentil, red/masoor dahl and tomato, Punjabi

25g vegetable ghee
½tsp chilli powder
½tsp cumin powder
½tsp dry ginger
¼tsp dry mustard
¼tsp turmeric

10g chopped garlic
160g mashed boiled red lentils/masoor dahl
90g chopped boiled onion
½tsp salt
250g skinned and chopped tomatoes

Fry spices in ghee gently for a couple of minutes add garlic, lentils, onion, salt and tomatoes. Boil for 1 minute.

Weight loss: 9%


### 124 Curry, lentil, red/masoor dahl and mung bean dahl

85g dry mung bean dahl
85g dry red lentils/masoor dahl
900ml water
90g chopped onion
1tsp salt

60g vegetable oil
½tsp cumin
1tsp turmeric
1tsp garam masala
1tsp chilli powder

Place mung dahl, lentils, water, onion and salt together in a pan. Bring to the boil, reduce heat and simmer for 30 minutes. In a separate pan heat oil, add spices and stir. Pour boiled mung dahl and lentils into the pan and simmer for 5 minutes.

Weight loss: 23%


### 125 Curry, lentil red/masoor dahl, mung bean dahl and tomato

28.6% boiled red lentils/masoor dahl
26.8% boiled mung bean dahl
26.3% water
6.9% tomato
5.3% onion

4.0% vegetable oil
0.8% green chilli
0.8% garlic
0.7% root ginger

Proportions are derived from dietary survey records.


### 126 Curry, lentil, red/masoor dahl, and vegetable

225g dry red lentils/masoor dahl
565ml water
33g vegetable oil
90g chopped onion
1tsp cumin
1tsp coriander

1tsp turmeric
½tsp ground ginger
170g chopped tomato
65g chopped carrot
160g sliced green pepper

Place lentils with water in a pan. Bring to the boil and simmer for 30 minutes. Heat the oil in a separate pan, add spices and heat, add tomatoes and cook through. Add remaining ingredients, mix well, cover and cook for 15 minutes. Stir vegetables into cooked lentils and serve.

Weight loss: 23%

### 127  Curry, lentil, whole/masoor, Gujerati

250g dry whole lentils/masoor
680ml water
150g chopped onion
60g butter
7g chopped green chilli
7g crushed root ginger
5g crushed garlic

2tsp salt
1tsp turmeric
55g tomato purée
1tsp garam masala
2tsp sugar
10ml lemon juice

Method as for curry, black gram, whole, Gujerati (No. 92).

Weight loss: 19%


### 128  Curry, lentil whole/masoor, Punjabi

170g dry whole lentils/masoor
900ml water
1½tsp salt
60g vegetable oil
90g chopped onion

½tsp cumin seeds
2tsp tomato purée
1tsp turmeric
1tsp chilli powder
2tsp garam masala

Bring lentils to the boil in salted water, reduce heat and simmer for 45 minutes. Heat the oil in a separate pan, add onion and cumin seeds, cook until golden brown. Add tomato purée and spices, simmer for 1 minute. Add the onion mixture to the lentil mix, simmer and serve.

Weight loss: 38%


### 129  Curry, mung bean dahl, Bengali

240g dry mung bean dahl
300ml water absorbed on soaking
1680ml water absorbed on boiling
60g butter
70g chopped onion

5g crushed root ginger
2g crushed garlic
4g chilli powder
4g mixed spices
7g salt

Soak mung dahl in water for 1 hour. Drain, add remaining ingredients and either simmer over a low heat until tender or cook in a pressure cooker for 10 minutes at 15lb pressure.

Weight loss: 33%


### 130  Curry, mung bean dahl, Punjabi

79.2% boiled mung bean dahl
10.4% water
6.1% onion
2.8% vegetable oil

0.6% green chilli
0.5% root ginger
0.4% garlic

Proportions are derived from dietary survey records.

### 131 Curry, mung bean dahl and spinach

| | |
|---|---|
| 115g dry mung bean dahl | 1tsp salt |
| 120g water absorbed on soaking | 2tsp chilli powder |
| 11g vegetable oil | 450g finely chopped spinach |
| 10g crushed garlic | 2tsp sugar |
| 1tsp turmeric | 11ml lemon juice |

Soak mung dahl for 1-2 hours. Heat oil, add garlic and fry, add salt, spices and sugar. Add mung dahl, mix well and put spinach on top. Cover and cook on a low heat until mung dahl are tender. Add lemon juice.

Weight loss: 21%


### 132 Curry, mung bean dahl and tomato

| | |
|---|---|
| 69.3% boiled mung bean dahl | 3.2% vegetable oil |
| 10.5% water | 0.8% root ginger |
| 8.6% tomatoes | 0.5% chilli powder |
| 6.8% onion | 0.2% garlic |

Proportions are derived from dietary survey records.


### 133 Curry, mung bean, whole, Gujerati

| | |
|---|---|
| 250g dry whole mung beans | 2tsp salt |
| 815ml water | 1tsp turmeric |
| 150g chopped onion | 55g tomato purée |
| 60g vegetable oil | 1½tsp sugar |
| 7g chopped green chilli | ½tsp coriander powder |
| 7g crushed root ginger | ½tsp cumin powder |
| 5g crushed garlic | 10ml lemon juice |

Wash and soak beans. Fry onion in the butter, add chilli, ginger and garlic. Add salt, turmeric, tomato purée and sugar and cook until the fat rises to the surface. Boil beans until soft. Add beans to the sauce and mix well. Add coriander, cumin and lemon juice and heat. If too thick add water and simmer.

Weight loss: 21%


### 134 Curry, mung bean, whole, Punjabi

| | |
|---|---|
| 220g dry whole mung beans | 40g butter |
| 90ml water, absorbed on soaking | 5g crushed root ginger |
| 1000ml water added during cooking | 2g crushed garlic |
| 120g chopped onion | 4g mixed spices |
| 100g chopped tomatoes | 7g salt |

Soak the beans. Drain, add remaining ingredients and either simmer over a low heat until tender or cook in a pressure cooker for 20 minutes at 15lb pressure.

Weight loss: 17%

## 135  Curry, mung bean, whole and turnip leaves

149g vegetable ghee
135g butter
35g chopped green chillies
26g crushed garlic
13g paprika
13g curry powder

13g chilli powder
18g salt
18g garam masala
900g boiled turnip leaves
618g boiled whole mung beans

Heat oil and fry chillies, garlic and spices. Add boiled leaves and mung beans, mix thoroughly.

Weight loss: 2%


## 136  Curry, okra

200g chopped onions
1tsp salt
57g vegetable oil
5g crushed garlic

10g root ginger
220g canned tomatoes
1tsp chilli powder
450g trimmed okra

Fry the onions and salt until brown. Add garlic, ginger, tomatoes and chilli powder until well blended and the oil starts to rise to the surface. Add okra and cook gently.

Weight loss: 26%


## 137  Curry, pea and potato

95g chopped onion
95g tomatoes
20g chopped root ginger
70g vegetable oil
165g diced potatoes

550ml water
5g salt
200g tinned peas
4g chilli powder
2g turmeric

Fry the onion, tomato and ginger in the oil. Add potatoes and water and simmer until cooked. Add remaining ingredients and simmer for a further 20 minutes.

Weight loss: 45%


## 138  Curry, pigeon pea dahl, with butter

240g dry pigeon pea dahl
250ml water absorbed on soaking
1680ml water added during cooking
60g butter
1tsp cumin seeds
1tsp mustard seeds
68g chopped onion
5g crushed root ginger

2g crushed garlic
4g chilli powder
2g cumin powder
2g coriander powder
7g salt
4tsp sugar
10ml lemon juice

Soak peas for 1 hour. Drain and cook in water until very soft. Liquidize. Heat the ghee, add cumin and mustard seeds and fry until they pop. Add to the liquidized peas followed by remaining ingredients. Simmer for 10-15 minutes.

Weight loss: 48%

### 139 Curry, pigeon pea dahl, with vegetable oil

240g dry pigeon pea dahl
250ml water absorbed on soaking
1680ml water added during cooking
60g vegetable oil
1tsp cumin seeds
1tsp mustard seeds
68g chopped onion
5g crushed root ginger

2g crushed garlic
4g chilli powder
2g cumin powder
2g coriander powder
7g salt
4tsp sugar
10ml lemon juice

Method as curry, pigeon pea dahl with butter (No. 138).

Weight loss: 48%


### 140 Curry, pigeon pea dahl and tomato

115g dry pigeon pea dahl
1800ml water
1tsp salt
22g vegetable oil
½tsp mustard seeds

85g chopped, skinned tomato
7g chopped green chilli
½tsp turmeric
½ tsp chilli powder
3tsp sugar

Place peas, water and salt together, bring to the boil, cover and simmer for 90 minutes. When cooked force through a coarse sieve. Heat the oil, add mustard seeds and then remaining ingredients, cook for 2 minutes. Add pea liquid to the pan and simmer for 20 minutes.

Weight loss: 45%


### 141 Curry, pigeon pea dahl with tomatoes and peanuts

250g dry pigeon pea dahl
815ml water
1tsp salt
11g vegetable oil
30g peanuts

100g chopped tomatoes
7g chopped green chilli
7g crushed root ginger
1tsp garam masala
½tsp turmeric

Soak peas for 2-3 hours in water. Add salt and cook until soft then liquidize. Heat oil, add peanuts then add tomatoes and cook until soft. Add all ingredients including liquidized peas and simmer for half an hour.

Weight loss: 25%


### 142 Curry, potato, Gujerati

30g vegetable oil
½tsp cumin seeds
½tsp black mustard seeds
¼tsp asafoetida
360g diced boiled potatoes
½tsp turmeric
½tsp cumin powder
½tsp coriander powder

1tsp salt
1tsp chilli powder
55g canned tomatoes
2tsp sugar
450ml water
10ml lemon juice
5g chopped coriander leaves

Heat oil and add cumin and mustard seed and asafoetida. After a few seconds add potatoes, spices, salt, tomatoes and sugar. Stir continuously and cook for 5 minutes. Add water, bring to boil, lower heat and simmer for 20 minutes. Add lemon juice and coriander leaves.

Weight loss: 35%

### 143 Curry, potato, Punjabi

| | |
|---|---|
| 90g chopped onion | 1tsp salt |
| 10g finely chopped green chillies | ½tsp chilli powder |
| 25g butter | 32g tomato purée |
| 1tsp turmeric | 240ml water |
| 1tsp ground ginger | 360g diced boiled potatoes |
| ½tsp coriander powder | |

Fry the onion and chillies in the butter until brown. Add turmeric, ginger, coriander, salt, chilli powder and tomato purée. Cook for 2 minutes. Add half the water together with the potatoes. Cook for 3 minutes. Add remaining water and boil, lower heat and simmer for 5 minutes.

Weight loss: 29%

### 144 Curry, potato and pea

| | |
|---|---|
| 450g chopped onions | 4g turmeric |
| 7g crushed garlic | 8g mixed spice |
| 30g vegetable oil | 360g diced boiled potatoes |
| 5g salt | 400ml boiling water |
| 4g chilli powder | 225g frozen peas |

Fry onion and garlic in the oil. Add spices and mix well. Add the potatoes and cook for 5 minutes. Add boiling water, cover and simmer for 10 minutes. Add frozen peas and simmer for a further 10 minutes.

Weight loss: 38%

### 145 Curry, red kidney bean, Gujerati

| | |
|---|---|
| 170g dry red kidney beans | ½tsp cumin seeds |
| 1200ml water | 90g chopped onion |
| 1tsp salt | 1tsp turmeric |
| 60g vegetable oil | 1tsp garam masala |
| 55g chopped tomatoes | 3tsp sugar |
| 14g chopped chillies | 30ml lemon juice |
| 20g crushed garlic | |

Boil the beans in salted water until soft. Heat the oil and cook tomatoes, chillies, garlic, cumin and onion. Add bean mix together with remaining ingredients, lower heat and simmer for 30 minutes.

Weight loss: 37%

### 146 Curry, red kidney bean, Punjabi

53.3% boiled red kidney beans
19.2% water
11.5% tomatoes
9.4% onions

5.3% vegetable oil
0.9% crushed root ginger
0.4% crushed garlic

Proportions are derived from dietary survey records.


### 147 Curry, red kidney and mung bean, whole

76.4% water
6.9% dry whole mung beans
6.5% dry red kidney beans

6.4% vegetable oil
3.8% butter

Proportions are derived from dietary survey records.

Weight loss: 20%


### 148 Curry, spinach

90g chopped onion
44g vegetable oil
400g canned tomatoes
½tsp salt
1tsp coriander powder

1tsp cumin powder
¼tsp turmeric
½tsp garlic paste
250g chopped spinach leaves

Fry onion in the oil until brown. Add tomatoes and the spices and fry for 5-10 minutes. Add spinach and mix well. Cook on a low heat until spinach is tender.

Weight loss: 35%


### 149 Curry, spinach and potato

200g chopped onions
1tsp salt
35g vegetable oil
5g crushed garlic
10g crushed root ginger

220g canned tomatoes
1tsp chilli powder
115g frozen spinach
50g diced potato
72ml water

Fry the onions and salt until brown. Add garlic, ginger, tomatoes and chilli powder, cook until well blended and the oil starts to rise to the surface. Add spinach, potato and water. Cook until potato is soft.

Weight loss: 37%


### 150 Curry, tinda and potato

325g onions
15g garlic
55g corn oil
50g root ginger

10g chilli powder
350g tomatoes
600g tinda gourd
370g potato

Quantities are derived from dietary survey records.

Weight loss: 15%

## 151 Curry, vegetable, frozen mixed vegetables

45g vegetable oil
½tsp cumin seeds
½tsp mustard seeds
90g finely chopped onion
10g tomato purée
1tsp turmeric powder
1tsp chilli powder

1tsp garam masala
1tsp salt
1tsp garlic powder
450g frozen mixed vegetables
100g diced potatoes
300ml water

Heat oil, add cumin seeds, mustard seeds and onion. Cook until onion is brown. Add remaining ingredients except water and stir continuously for a further 5 minutes. Add water to the vegetables and boil, reduce heat and simmer for 20 minutes.

Weight loss: 20%


## 152 Curry, vegetable, in sweet sauce

33.4% sweet curry sauce
23.8% canned tomatoes
10.7% boiled carrots

10.7% boiled potatoes
10.7% boiled cauliflower
10.7% boiled courgette

Proportions are derived from dietary survey records.


## 153 Curry, vegetable, Islami

22g vegetable oil
½tsp mustard seeds
¼tsp fenugreek seeds
¼tsp cumin seeds
125g diced potatoes
125g cauliflower florets
125g diced aubergine

125g chopped onion
220g canned tomatoes
1tsp garlic paste
22g tomato purée
1tsp coriander powder
1tsp cumin powder
¼tsp turmeric

Heat oil, add seeds and fry until they start to pop. Add mixed vegetables and stir. Add tomatoes and cook through. Add remaining ingredients, reduce heat and cook until the sauce thickens and vegetables are cooked.

Weight loss: 25%


## 154 Curry, vegetable, Pakistani

200g diced potatoes
180g diced carrots
50g diced turnip
115g peas
115g chopped runner beans
370ml water
90g finely chopped onion
11g corn oil
2g crushed garlic

15g coriander powder
1tsp turmeric
1tsp ground ginger
1tsp chilli powder
½tsp cumin powder
44g tomato purée
10g desiccated coconut
10ml lemon juice

Parboil the vegetables, drain and reserve the water. Fry onion in oil until brown, add garlic and spices and continue for 2 minutes. Add tomato purée, some of the vegetable water, boil, lower heat and simmer for 10 minutes. Stir in coconut and vegetables. Add lemon juice and simmer until tender.

Weight loss: 24%

## 157 Curry, vegetable, West Indian

22g vegetable oil
2g crushed garlic
10g curry powder
260g sliced and salted aubergine
225g trimmed okra

225g chopped spinach
50g diced potato
5ml lime juice
½tsp salt

Fry garlic in oil, stir in curry powder. Add vegetables, lime juice and salt. Stir, cover and steam for 15-20 minutes.

Weight loss: 11%


## 158 Curry, vegetable, with yogurt

180g sliced onions
44g vegetable oil
2tsp coriander powder
2tsp turmeric
1tsp curry powder
10g chopped root ginger
5g crushed garlic

180g sliced carrots
350g sliced courgettes
300ml water
½tsp salt
400g cauliflower florets
150g low fat natural yogurt

Fry onions in oil until soft. Add spices and garlic and continue to cook for a minute. Add carrots and courgettes and cook for 2-3 minutes. Add water and seasoning, cover and simmer for 10 minutes. Add cauliflower and cook for 10 minutes. Stir in yogurt and heat through.

Weight loss: 21%


## 159 Dal Dhokari

Pastry:
250g wholemeal flour
30g chick pea flour
30g turmeric powder
1tsp salt
½tsp chilli powder
200ml water

Sauce:
250g split pigeon peas
1140ml water
30g vegetable oil
30g peanuts
5g crushed green chilli
5g crushed root ginger
2tsp salt
½tsp turmeric
10ml lemon juice
1tsp garam masala
2tsp sugar
5g chopped coriander leaves

Mix pastry ingredients to a smooth dough. Leave to rest for at least 30 minutes. Wash and soak peas for 2-3 hours then cook until soft. Liquidize and add water. Heat oil, add peanuts, liquidized peas, chilli, ginger, salt and turmeric. Roll the dough very thin and cut into small squares. Drop these into the boiling pea mixture, cover and cook at low temperature. When pastry has cooked add lemon juice, garam masala, sugar and coriander leaves and serve.

Weight loss: 12%

## 160 Dosa, plain

170g dry black gram dahl
55g raw white rice
560ml water
¼tsp bicarbonate of soda

1tsp chilli powder
1tsp salt
vegetable oil absorbed on frying (40g)

Soak dahl and rice in water overnight. Liquidize to pouring consistency. Add remaining ingredients, mix well and stand for 15 minutes. Cook like a conventional pancake but very thin.

Weight loss: 38%


## 161 Dosa, filling, vegetable

25g vegetable oil
2tsp mustard seeds
1tsp salt
1tsp turmeric

¼tsp chilli powder
140g boiled potatoes, cut into cubes
180g chopped onion
225g chopped tomatoes

Heat oil, add mustard seeds, curry leaves and spices and cook for 2 minutes. Add potatoes, onion and tomatoes and simmer until potatoes are soft. Cook until mixture is dry. Use as filling for Dosa plain (No. 160).

Weight loss: 23%


## 162 Falafel, fried in vegetable oil

100g dry chick peas
203ml water absorbed on soaking
130g finely chopped onions
6g crushed garlic
20g finely chopped parsley

1½tsp cumin powder
1½tsp coriander powder
½tsp baking powder
½tsp salt
vegetable oil absorbed on frying (37g)

Soak chick peas overnight and drain. Mince or grind in a processor. Add peas to remaining ingredients and blend into a very smooth paste. Form into small balls and leave for 15 minutes. Deep fry in hot oil.

Weight loss: 22%


## 163 Flan, broccoli

25% flour
20% milk
15% egg
15% broccoli
10% margarine

10% cheese
2% UHT cream
2% tomato purée
1% mustard

Proportions are derived from dietary survey records.


## 164 Flan, broccoli, wholemeal

25% wholemeal flour
20% milk
15% egg
15% broccoli
10% margarine

10% cheese
2% UHT cream
2% tomato purée
1% mustard

Proportions are derived from dietary survey records.

### 165 Flan, cauliflower cheese

| | |
|---|---|
| 28.8% milk | 1.9% tomato |
| 28.8% boiled cauliflower | 1% butter |
| 19.2% flour | 1% egg |
| 9.6% margarine | 1% boiled onion |
| 4.8% cheese | 1% salt |
| 2.9% single cream | |

Proportions are derived from dietary survey records.

### 166 Flan, cauliflower cheese, wholemeal

| | |
|---|---|
| 28.8% milk | 1.9% tomato |
| 28.8% boiled cauliflower | 1% butter |
| 19.2% wholemeal flour | 1% egg |
| 9.6% margarine | 1% boiled onion |
| 4.8% cheese | 1% salt |
| 2.9% single cream | |

Proportions are derived from dietary survey records.

### 167 Flan, cheese and mushroom

| | |
|---|---|
| 200g shortcrust pastry | 100g cheese |
| 120g egg | 200g milk |
| 25g boiled onion | ½tsp salt |
| 100g boiled mushroom | |

Quantities are derived from dietary survey records.

Weight loss: 25%

### 168 Flan, cheese and mushroom, wholemeal

| | |
|---|---|
| 200g wholemeal shortcrust pastry | 100g cheese |
| 120g egg | 200g milk |
| 25g boiled onion | ½tsp salt |
| 100g boiled mushroom | |

Quantities are derived from dietary survey records.

Weight loss: 25%

### 169 Flan, cheese, onion and potato

| | |
|---|---|
| 120g boiled potatoes | 60g boiled onion |
| 5g margarine | 200g shortcrust pastry |
| 240g cheese | |

Quantities are derived from dietary survey records.

Weight loss: 8%

### 170 Flan, cheese, onion and potato, wholemeal

| | |
|---|---|
| 120g boiled potatoes | 60g boiled onion |
| 5g margarine | 200g wholemeal shortcrust pastry |
| 240g cheese | |

Quantities are derived from dietary survey records.

Weight loss: 8%

## 171   Flan, lentil and tomato

Pastry:
100g flour                        30ml cold water
25g margarine

Filling:
100g chopped onion           275ml water
30g chopped celery           5g chopped fresh parsley
1tsp vegetable oil            35g grated cheese
125g dry red lentils         2tsp sesame seeds
200g canned tomatoes

Fry the onion and celery in oil until soft. Add lentils, tomatoes and water and simmer until the lentils are fully cooked. Prepare pastry, line a 7inch/18cm flan dish and bake blind. Add parsley to filling mixture which should have quite a dry consistency. Pour filling into flan case, sprinkle with cheese and sesame seeds. Bake in oven at 190°C/mark 5 for 15 minutes until cheese melts.

Weight loss: 24%

## 172   Flan, lentil and tomato, wholemeal

Pastry:
100g wholemeal flour        30ml cold water
25g margarine

Filling:
100g chopped onion           275ml water
30g chopped celery           5g chopped fresh parsley
1tsp vegetable oil            35g grated cheese
125g dry red lentils         2tsp sesame seeds
200g canned tomatoes

Method as flan, lentil and tomato (No. 171).

Weight loss: 24%

## 173   Flan, spinach

90g chopped onion           2 eggs
22g vegetable oil            50g Parmesan cheese
350g drained and chopped   ½tsp salt
  frozen spinach          ½tsp pepper
250g cottage cheese       225g raw shortcrust pastry

Fry the onion in oil until soft. Add spinach and cook through. Cool slightly and mix in cheeses, eggs and seasoning. Line a flan dish with pastry. Pour filling mixture into dish and bake at 200°C/mark 6 for 30-35 minutes.

Weight loss: 16%

### 174 Flan, spinach, wholemeal

| | |
|---|---|
| 90g chopped onion | 2 eggs |
| 22g vegetable oil | 50g Parmesan cheese |
| 350g drained and chopped | ½tsp salt |
|   frozen spinach | ½tsp pepper |
| 250g cottage cheese | 225g raw wholemeal shortcrust pastry |

Method as flan, spinach (No. 173).

Weight loss: 16%


### 175 Flan, vegetable

| | |
|---|---|
| 30.8% shortcrust pastry | 15.4% boiled onion |
| 15.4% boiled carrot | 15.4% white sauce made with skimmed milk |
| 15.4% boiled broccoli | 7.7% cheese |

Proportions are derived from dietary survey records.


### 176 Flan, vegetable, wholemeal

| | |
|---|---|
| 30.8% wholemeal shortcrust pastry | 15.4% boiled onion |
| 15.4% boiled carrot | 15.4% white sauce made with skimmed milk |
| 15.4% boiled broccoli | 7.7% cheese |

Proportions are derived from dietary survey records.


### 177 Fu-fu, sweet potato

756g boiled sweet potato(salted water) 75g rice flour

Pound sweet potatoes and flour together until well mixed and smooth.


### 178 Fu-fu, yam

| | |
|---|---|
| 730g boiled yam(salted water) | 75g rice flour |

Pound yam and flour together until well mixed and smooth.


### 179 Garlic mushrooms

250g mushrooms
2g garlic
40g butter

Clean mushrooms and remove stems. Crush the garlic and sauté in butter. Fill mushroom caps with the garlic butter mixture and grill for 5-7 minutes.

Weight loss: 19%


### 180 Guacamole

| | |
|---|---|
| 195g avocado flesh | 85g tomato |
| 20ml lemon juice | 1g salt |

Mash avocado and sprinkle with lemon juice. Add finely chopped tomato and salt.

### 181 Khadhi

| | |
|---|---|
| 50g chick pea flour | 3tsp salt |
| 500g natural yogurt | ½tsp turmeric |
| 850ml water | 3tsp sugar |
| 7g chopped green chilli | 15g butter ghee |
| 7g crushed root ginger | 3g cumin seeds |
| 5g crushed garlic | 5g finely chopped coriander leaves |

Mix flour, yogurt and water and liquidize. Add chilli, ginger, garlic, turmeric and sugar. Heat ghee, add cumin seeds, pour in yogurt and stir until it starts to boil. Simmer for 15-20 minutes. Add curry and coriander leaves.

Weight loss: 20%


### 182 Khatiyu

| | |
|---|---|
| 50g vegetable oil | ½tsp turmeric |
| 50g chick pea flour | ½tsp chilli powder |
| 675g boiled whole pigeon peas | 30g jaggary |
| 20g tamarind pulp | 6g salt |

Heat oil, add chick pea flour and cook until brown. Add remaining ingredients and simmer.

Weight loss: 3%


### 183 Khichadi, with butter ghee

| | |
|---|---|
| 50g butter ghee | 150g dry whole mung beans |
| 3tsp salt | 1145ml water |
| 1tsp turmeric | 300g raw white rice |

Heat ghee, add salt, turmeric, mung beans and water. Cook mung beans until half cooked, add rice and mix well. Cook until rice is soft and all water is absorbed.

Weight loss: 28%

### 184 Khichadi, with vegetable oil

| | |
|---|---|
| 50g vegetable oil | 150g dry whole mung beans |
| 3tsp salt | 1145ml water |
| 1tsp turmeric | 300g raw white rice |

Method as khichadi, with butter ghee (No. 183).

Weight loss: 28%


### 185 Lasagne, spinach

| | |
|---|---|
| 500g boiled lasagne | 250g canned tomatoes |
| 250g boiled spinach | 184g cheese sauce |

Quantities are derived from dietary survey records.

Weight loss: 8%

### 186 Lasagne, spinach, wholemeal

As lasagne, spinach (No. 185) except using wholemeal pasta and wholemeal flour.

Weight loss: 8%

### 187  Lasagne, vegetable

565g boiled lasagne
50g boiled spinach
80g boiled carrot
40g boiled pepper
50g boiled courgette
64g boiled mushroom
60g boiled celery

24g margarine
420g canned tomatoes
18g tomato ketchup
46g tomato purée
350ml milk
100g cheese
54g flour

Quantities are derived from dietary survey records. Ingredients baked at 200°C/mark 6 for 35 minutes.

Weight loss: 12%


### 188  Lasagne, vegetable, wholemeal

As lasagne, vegetable (No. 187) using wholemeal pasta and flour.

Weight loss: 12%


### 191  Leeks in cheese sauce, made with whole milk

58% boiled leeks
40% cheese sauce made with whole milk
2% cheese

Proportions are derived from a recipe review.


### 192  Leeks in cheese sauce, made with semi-skimmed milk

58% boiled leeks
40% cheese sauce made with semi-skimmed milk
2% cheese

Proportions are derived from a recipe review.


### 193  Leeks in cheese sauce, made with skimmed milk

58% boiled leeks
40% cheese sauce made with skimmed milk
2% cheese

Proportions are derived from a recipe review.

## 194  Lentil cutlets, fried in vegetable oil

112g dry red lentils
120g water absorbed on soaking
70g mashed potato
10g chopped green chillies
5g crushed root ginger
5g chopped coriander leaves
7g salt
30ml lemon juice
2tsp garam masala
10g packet breadcrumbs
vegetable oil absorbed on frying (27g)

Soak lentils for at least 5-6 hours. Grind soaked lentils into a thick paste, add mashed potato and remaining ingredients except breadcrumbs. Mix well. Shape into small cutlets, coat with breadcrumbs and shallow fry.

Weight loss: 9%


## 195  Lentil pie

235g boiled whole or red lentils
90g chopped onion
100g grated carrot
15g vegetable oil
15g wholemeal flour
10ml lemon juice
10g chopped fresh parsley
80g wholemeal breadcrumbs

Boil and drain lentils. Fry the onion and carrots in the oil until soft, add flour and stir. Mix with lentils, lemon juice and parsley. Put into pie dish and cover with breadcrumbs. Bake at 180°C/mark 4 for 15 minutes.

Weight loss: 20%


## 196  Lentil and cheese pie

540g boiled whole or red lentils
22g vegetable oil
120g chopped onion
1tsp marmite
120g cheese

Boil and drain lentils. Fry onion in the oil until soft. Mix with lentils and marmite. Add half the cheese and put into a shallow pie dish. Sprinkle with remaining cheese. Bake at 180°C/mark 4 for 15 minutes.

Weight loss: 11%


## 197  Lentil and potato pie

350g boiled whole or red lentils
350g boiled potato
15g margarine
90g chopped onion
1tsp salt

Mix together lentils and mashed boiled potato. Fry onion in margarine and add to lentil mixture together with salt. Bake at 180°C/mark 4 for approximately 20 minutes until browned.

Weight loss: 16%

### 198 Lentil roast

90g chopped onion
45g grated carrot
11g vegetable oil
350g boiled whole or red lentils

100g wholemeal breadcrumbs
15g soya sauce
10g chopped fresh parsley

Fry the onion and carrot in the oil. Add mashed lentils, beaten egg and remaining ingredients. Pack into an oiled loaf tin and cover with foil. Bake at 190°C/mark 5 for 30-40 minutes.

Weight loss: 11%


### 199 Lentil roast, with egg

90g chopped onion
45g grated carrot
11g vegetable oil
350g boiled whole or red lentils

1 egg
100g wholemeal breadcrumbs
15g soya sauce
10g chopped fresh parsley

Method as lentil roast (No. 198). Add beaten egg with lentils.

Weight loss: 11%


### 200 Lentil and nut roast

90g chopped onion
45g grated carrot
11g vegetable oil
410g boiled whole or red lentils

75g wholemeal breadcrumbs
125g chopped mixed nuts
15g soya sauce

Fry the onion and carrot in the oil. Add mashed lentils and remaining ingredients. Pack into an oiled loaf tin and cover with foil. Bake at 190°C/mark 5 for 30-40 minutes.

Weight loss: 11%


### 201 Lentil and nut roast, with egg

90g chopped onion
45g grated carrot
11g vegetable oil
410g boiled whole or red lentils

1 egg
75g wholemeal breadcrumbs
125g chopped mixed nuts
15g soya sauce

Method as lentil and nut roast (No. 200). Add beaten egg with lentils.

Weight loss: 11%


### 202 Lentil and rice roast

90g chopped onion
11g vegetable oil
20g wholemeal flour
45ml water

650g boiled brown rice
350g boiled whole or red lentils
24g soya sauce

Fry the onion in the oil, add flour and water and mix. Mix with the rice, mashed lentils and soy sauce. Pack into an oiled loaf tin and cover with foil. Bake at 190°C/mark 5 for 30-40 minutes.

Weight loss: 8%

### 203 Lentil and rice roast, with egg

90g chopped onion
11g vegetable oil
20g wholemeal flour
45ml water

650g boiled brown rice
350g cooked whole or red lentils
1 egg
24g soya sauce

Recipe as lentil and rice roast (No. 202). Add egg with rice and lentils.

Weight loss: 8%

### 204 Mchicha

450g chopped spinach
14g vegetable oil
60g chopped onion
55g chopped tomato

½tsp salt
½tsp pepper
9g curry powder

Steam the spinach for a few minutes. Heat the oil and fry onion and tomato gently until tender. Add spinach and seasoning. Stir well.

Weight loss: 33%

### 205 Moussaka, vegetable

500g aubergines
120g vegetable oil
225g finely chopped onion
5g crushed garlic
225g chopped mushrooms
280g cooked red lentils
25g tomato purée
250g sliced tomatoes

500g boiled potatoes
½tsp salt
½tsp pepper
1tsp mixed herbs
500ml white sauce
2 eggs
150g grated cheese

Lightly fry thinly sliced aubergines and drain. Fry onion and garlic. Add mushrooms and lentils, season. Cook for 3 minutes and pour into a dish. Cover with aubergines then tomatoes and finally sliced boiled potatoes. Beat eggs into white sauce and pour over potatoes. Sprinkle with cheese. Bake for 20-30 minutes at 180°C/mark 4.

Weight loss: 12%

### 207 Mushroom Dopiaza, retail

mushroom
onion
tomato

water
rapeseed oil
mixed spices

Proportions obtained from a manufacturer.

Weight loss: 5% on re-heating

### 208 Nut croquettes, fried in sunflower oil

11g sunflower oil
90g chopped onion
5g crushed garlic
75g chopped mushrooms
1tsp ground coriander
1tsp cumin

40g flour
150ml water
125g chopped mixed nuts
125g wholemeal breadcrumbs
8g flour for coating
sunflower oil absorbed on frying (57g)

Heat oil and fry onion, garlic and mushrooms. Add spices and flour, remove from heat and stir in water. Add remaining ingredients and mix well. Shape, coat in flour and shallow fry until brown and crisp.

Weight loss: 22%


### 209 Nut croquettes, fried in vegetable oil

11g vegetable oil
90g chopped onion
5g crushed garlic
75g chopped mushrooms
1tsp ground coriander
1tsp cumin

40g flour
150ml water
125g chopped mixed nuts
125g wholemeal breadcrumbs
8g flour for coating
vegetable oil absorbed on frying (57g)

Method as nut croquettes, fried in sunflower oil (No. 208).

Weight loss: 22%


### 210 Nut cutlets, retail, fried in sunflower oil

boiled white rice
water
boiled brown rice
almonds
sesame seeds
breadcrumbs
carrots
sunflower oil absorbed on frying (9.5%)

mushrooms
potato
rapeseed oil
onions
cornflour
chives
salt

Proportions are obtained from a manufacturer.

Weight loss: 12%


### 211 Nut cutlets, retail, fried in vegetable oil

boiled white rice
water
boiled brown rice
almonds
sesame seeds
breadcrumbs
carrots
vegetable oil absorbed on frying (9.5%)

mushrooms
potato
rapeseed oil
onions
cornflour
chives
salt

Proportions are obtained from a manufacturer.

Weight loss: 12%

## 212  Nut cutlets, retail, grilled

boiled white rice
water
boiled brown rice
almonds
sesame seeds
breadcrumbs
carrots

mushrooms
potato
rapeseed oil
onions
cornflour
chives
salt

Proportions obtained from a manufacturer.


## 213  Nut roast

90g chopped onion
11g vegetable oil
20g flour
140ml water

225g chopped mixed nuts
115g wholemeal breadcrumbs
1tsp marmite
1tsp mixed herbs

Fry onion in the oil. Add flour and water and thicken. Mix in nuts, breadcrumbs, marmite and herbs. Pack into a loaf tin and cover with foil. Bake at 190°C/mark 5 for 35-45 minutes.

Weight loss: 13%


## 214  Nut roast, with egg

90g chopped onion
11g vegetable oil
20g flour
140ml water

225g chopped mixed nuts
115g wholemeal breadcrumbs
1tsp marmite
1tsp mixed herbs
1 beaten egg

Method as nut roast (No. 213). Add beaten egg with dry ingredients.

Weight loss: 13%


## 215  Nut and rice roast

90g chopped onion
11g vegetable oil
100g wholemeal breadcrumbs
225g chopped mixed nuts

250g boiled brown rice
1tsp marmite
1tsp mixed herbs
½tsp salt

Fry the onion in the oil. Mix all ingredients together. Pack into a loaf tin and cover with foil. Bake at 190°C/mark 5 for 35-45 minutes.

Weight loss: 11%


## 216  Nut and rice roast, with egg

90g chopped onion
11g vegetable oil
100g wholemeal breadcrumbs
225g chopped mixed nuts

250g boiled brown rice
1tsp marmite
1tsp mixed herbs
½tsp salt
1 beaten egg

Method as nut and rice roast (No. 215). Add beaten egg with dry ingredients.

Weight loss: 11%

217 **Nut and seed roast**

| | |
|---|---|
| 90g chopped onion | 65g sunflower seeds |
| 11g vegetable oil | 115g wholemeal breadcrumbs |
| 20g flour | 1tsp marmite |
| 140ml water | 1tsp mixed herbs |
| 160g chopped mixed nuts | |

Fry onion in the oil. Add flour and water and thicken. Mix in nuts, breadcrumbs, marmite and herbs. Pack into a loaf tin and cover with foil. Bake at 190°C/mark 5 for 35-45 minutes.

Weight loss: 13%


218 **Nut and seed roast, with egg**

| | |
|---|---|
| 90g chopped onion | 65g sunflower seeds |
| 11g vegetable oil | 115g wholemeal breadcrumbs |
| 20g flour | 1tsp marmite |
| 140ml water | 1tsp mixed herbs |
| 160g chopped mixed nuts | 1 beaten egg |

Method as nut and seed roast (No. 217). Add beaten egg with dry ingredients.

Weight loss: 13%


219 **Nut and vegetable roast**

| | |
|---|---|
| 90g chopped onion | 1tsp marmite |
| 60g grated carrot | 20g tomato purée |
| 30g chopped celery | 200g chopped mixed nuts |
| 22g vegetable oil | 120g wholemeal breadcrumbs |
| 40g flour | 1tsp mixed herbs |
| 100ml water | |

Fry the onion, carrots and celery in the oil. Add flour, water, marmite and tomato purée and thicken. Add remaining ingredients and mix well. Pack into a loaf tin and press. Cover with foil and bake at 190°C/mark 5 for 35-40 minutes

Weight loss: 11%


220 **Nut and vegetable roast, with egg**

| | |
|---|---|
| 90g chopped onion | 1tsp marmite |
| 60g grated carrot | 20g tomato purée |
| 30g chopped celery | 200g chopped mixed nuts |
| 22g vegetable oil | 120g wholemeal breadcrumbs |
| 40g flour | 1tsp mixed herbs |
| 100ml water | 1 beaten egg |

Method as nut and vegetable (No. 219). Add beaten egg with dry ingredients.

Weight loss: 11%

### 221　Nut, mushroom and rice roast

80g grated carrots
90g chopped onion
100g chopped mushrooms
35g vegetable oil
375g boiled brown rice
150g chopped mixed nuts

15g soya sauce
100g wholemeal breadcrumbs
½tsp mixed herbs
1tsp marmite
140ml water

Fry carrots, onion and mushroom in the oil. Add rice, nuts, soy sauce, herbs and breadcrumbs, mix well. Dissolve the marmite in water and mix into other ingredients. Pour into a loaf tin and press. Bake at 190°C/mark 5 for 40 minutes.

Weight loss: 26%

### 222　Okra with tomatoes and onion, Greek

150g chopped onion
125g olive oil
225g chopped tomatoes
700g trimmed okra

½tsp pepper
½tsp vinegar
½tsp salt

Fry onions in oil until soft. Add tomatoes, okra and seasoning, reduce heat and simmer for 20-30 minutes.

Weight loss: 34%

### 223　Okra with tomatoes and onion, West Indian

90g thinly sliced onion
5g garlic
225g peeled and chopped tomatoes
2tsp salt
½tsp pepper

1tsp chilli powder
½tsp cumin
700g trimmed okra
50g vegetable oil

Fry onion and garlic until soft. Add tomatoes and seasoning and continue to fry for 3 minutes. Stir in okra, reduce heat and simmer for 20-30 minutes.

Weight loss: 30%

### 224　Pakora/bhajia, aubergine, fried in vegetable oil

240g chick pea flour
1tsp salt
1tsp chilli powder

210ml water
520g sliced aubergine
vegetable oil absorbed on frying (155g)

Make a batter with the flour, ½tsp of the salt and chilli powder and water. Sprinkle aubergines with remaining salt and chilli powder. Heat the cooking oil, dip each aubergine piece into the batter and place in the hot oil. Deep fry until golden brown.

Weight loss: 33%

### 225　Pakora/bhajia, cauliflower, fried in vegetable oil

240g chick pea flour
1tsp salt
1tsp chilli powder

210ml water
500g small cauliflower florets
vegetable oil absorbed on frying (160g)

Method as for pakora/bhajia, aubergine, fried in vegetable oil (No. 224).

Weight loss: 34%

### 226 Pakora/bhajia, onion, fried in vegetable oil

225g chick pea flour
½tsp salt
½tsp chilli powder
½tsp garam masala

10g finely chopped coriander leaves
200ml water
180g onion slices
vegetable oil absorbed in frying (58g)

Mix all ingredients together to a thick consistency. Drop tablespoonsful of the mixture into hot oil and deep fry until golden brown.

Weight loss: 29%


### 228 Pakora/bhajia, potato, fried in vegetable oil

185g chick pea flour
1tsp salt
1tsp chilli powder

165ml water
350g peeled, thinly sliced potatoes
vegetable oil absorbed on frying (94g)

Method as for pakora/bhajia, aubergine, fried in vegetable oil (No. 224).

Weight loss: 33%


### 229 Pakora/bhajia, potato and cauliflower, fried in vegetable oil

200g chick pea flour
16g melon seeds
2tsp salt
2tsp garam masala
180ml water

156g diced potatoes
80g chopped onions
100g chopped cauliflower
vegetable oil absorbed on frying (129g)

Make a batter with the flour, seeds, 1tsp each of salt and garam masala and water. Sprinkle vegetables with remaining salt and garam masala. Heat the cooking oil and drop spoonsful of the mixture into the fat. Deep fry until golden brown.

Weight loss: 23%


### 230 Pakora/bhajia, potato, carrot and pea, fried in vegetable oil

450g chick pea flour
1tsp salt
1tsp chilli powder
395ml water

100g diced potato
55g diced carrot
40g peas
vegetable oil absorbed on frying (170g)

Method as pakora/bhajia, aubergine, fried in vegetable oil (No. 224).

Weight loss: 29%


### 231 Pakora/bhajia, spinach, fried in vegetable oil

155g chick pea
1tsp salt
1tsp chilli powder

125ml water
80g finely chopped spinach
vegetable oil absorbed on frying (48g)

Mix all ingredients together to a thick consistency. Drop teaspoonsful of the mixture into hot oil and deep fry until golden brown.

Weight loss: 37%

232 **Pakora/bhajia, vegetable, retail**

| | |
|---|---|
| potato | cauliflower |
| onion | spinach |
| chick pea flour | self-raising flour |
| rapeseed oil | mixed spices |
| water | |

Proportions obtained from a manufacturer.


233 **Pancakes, stuffed with vegetables**

320g prepared pancakes
Filling:

| | |
|---|---|
| 50g chopped mushrooms | 90g chopped onion |
| 200g canned tomatoes | 1tsp mixed herbs |

Prepare the filling by cooking all the ingredients for approximately 15 minutes. Fill and roll up pancakes. Place under a grill to reheat if necessary.

Weight loss: 20% for filling


234 **Pancakes, stuffed with vegetables, wholemeal**

320g prepared wholemeal pancakes
Filling:

| | |
|---|---|
| 50g chopped mushrooms | 90g chopped onion |
| 200g canned tomatoes | 1tsp mixed herbs |

Method as pancakes, stuffed with vegetables (No. 233).

Weight loss: 20% for filling


235 **Pastichio**

| | |
|---|---|
| 90g chopped onion | ½tsp oregano |
| 55g butter | ½tsp cinnamon |
| 10g crushed garlic | 1tsp salt |
| 160g chopped green pepper | ½tsp pepper |
| 115g sliced mushrooms | 410g boiled macaroni |
| 225g dry red lentils | 430g white sauce |
| 400g canned tomatoes | 2 eggs |
| 568ml water | 85g grated cheese |

Fry onions in butter until soft, add garlic, pepper and mushrooms and cook for 5 minutes. Add remaining ingredients (except macaroni), bring to the boil and simmer for 45 minutes. Place macaroni in a pie dish and spoon the lentil mix over the top. Beat eggs into the white sauce and pour over dish mixture. Sprinkle with cheese and bake at 200°C/mark 6 for 40-50 minutes.

Weight loss: 30% preparing lentil mix

        10% baking dish

### 236  Pasty, vegetable

50% cooked shortcrust pastry
15% boiled potato
7.5% water
6.3% boiled carrot

6.3% boiled parsnip
6.3% boiled onion
6.3% boiled cabbage
2.5% flour

Proportions are derived from dietary survey records.


### 237  Pasty, vegetable, wholemeal

50% cooked wholemeal shortcrust
 pastry
15% boiled potato
7.5% water
6.3% boiled carrot

6.3% boiled parsnip
6.3% boiled onion
6.3% boiled cabbage
2.5% flour

Proportions are derived from dietary survey records.


### 238  Peppers, stuffed with rice

2 green peppers(380g)
40g chopped onion
10g vegetable oil
85g skinned and chopped tomatoes

30g raisins
170g boiled white rice
5g chopped fresh parsley
½tsp salt

Cut top from peppers and scoop out seeds and membranes. Parboil for 2-3 minutes and drain thoroughly. Fry the onions, remove from heat and mix with remaining ingredients. Pile filling into peppers. Cover tops with foil and bake at 200°C/mark 6 for 20-30 minutes.

Weight loss: 12%


### 239  Peppers, stuffed with vegetables, cheese topping

2 green peppers(380g)
33g vegetable oil
90g chopped onion
150g skinned and chopped tomatoes
100g sliced mushrooms

200g canned sweetcorn
1tsp salt
1tsp mixed herbs
20g tomato purée
50g grated cheese

Cut tops from peppers and scoop out seeds and membranes. Parboil for 2-3 minutes and drain thoroughly. Fry the onions until soft, add tomato and mushrooms and continue. Add remaining ingredients except cheese and simmer until reduced. Pile into peppers, sprinkle cheese on top and place in oven and bake at 200°C/mark 6 for 20-30 minutes.

Weight loss: 18% preparation of stuffing

   5% baking peppers

## 240 Pesto sauce

40g fresh basil
50g pine nuts
125g grated parmesan cheese
10g crushed garlic
60g olive oil

Put all ingredients except oil in a blender and mix. Slowly add oil and mix until well incorporated.


## 241 Pie, Quorn and vegetable

90g chopped onion
25g margarine
250g Quorn
25g flour
140ml water
284ml semi-skimmed milk
75g sweetcorn kernels
75g strips of carrots
50g cheese
½tsp salt
225g puff pastry
½ beaten egg

Fry the onion in margarine until soft, add Quorn and fry for a further 2-3 minutes. Add flour and cook for 1 minute stirring continuously. Gradually add milk and water, bring to the boil. Add sweetcorn and carrots, stir in grated cheese and allow to melt, season. Pour into ½ pint pie dish. Cover with pastry brushed with beaten egg. Bake at 200°C/mark 6 for 20-25 minutes.

Weight loss: 19%


## 242 Pie, spinach

675g sliced leeks, white part only
100g olive oil
1400g chopped spinach
70g spring sliced onion
5g fresh dill
450g feta cheese
240g beaten eggs(approx. 5)
100ml evaporated milk
450g flaky pastry

Blanch leeks then fry in a little oil until transparent. Add spinach, onions and dill. Break up feta cheese and mix with the vegetables. Add eggs, milk and remaining oil. Line a tin with half the pastry, spread the filling over the pastry base. Cover with the remaining pastry and tuck in well. Bake at 190°C/mark 5 for 45 minutes.

Weight loss: 23%


## 243 Pie, vegetable

100g chopped onion
100g sliced carrot
100g sliced courgettes
60g chopped celery
50g sliced mushrooms
80g chopped red pepper
100g potatoes
200g canned tomatoes
100ml water
2tsp cornflour
1tsp mixed herbs
1tsp marmite
300g raw shortcrust pastry

Place vegetables in a pan, together with herbs and marmite. Bring to the boil and simmer for 20-25 minutes. Make cornflour into a paste, add to pan, boil and stir until mixture thickens. Pour into pie dish and leave to cool. Roll pastry to fit dish size. Cut an additional 1 inch strip from remaining pastry, wet and place around the edge of the dish. Cover with pastry top and seal edges. Bake at 200°C/mark 6 for 30-40 minutes.

Weight loss: 15%

### 244 Pie, vegetable, wholemeal

100g chopped onion
100g sliced carrot
100g sliced courgettes
60g chopped celery
50g sliced mushrooms
80g chopped red pepper
100g potatoes

200g canned tomatoes
100ml water
2tsp cornflour
1tsp mixed herbs
1tsp marmite
300g raw wholemeal shortcrust pastry

Method as pie, vegetable (No. 243).

Weight loss: 15%

### 245 Pilaf, rice with spinach

90g finely chopped onion
15g corn oil
900g chopped spinach
225g raw white rice

568ml water
5g salt
2g pepper

Fry onion in the oil, add spinach and cook for 5 minutes until softened. Add remaining ingredients, cover and cook for 15-20 minutes or until the rice is soft. Remove lid, cover with a thick cloth, replace lid and leave for 10 minutes so the cloth absorbs the steam.

Weight loss: 33%

### 246 Pilaf, rice with tomato

90g finely chopped onion
30g corn oil
450g raw white rice
60ml lemon juice

225g chopped tomatoes
900ml stock
7g salt
5g pepper

Fry onion in the oil, add rice and stir. Add remaining ingredients, cover and cook for 15-20 minutes or until the rice is soft. Remove lid, cover with a thick cloth, replace lid and leave for 10 minutes so the cloth absorbs the steam.

Weight loss: 19%

### 247 Pilau, egg and potato

15.6% boiled egg
51.3% boiled white rice
12.5% onions fried in corn oil
15.6% boiled potatoes

3.1% green chillies
1.3% salt
0.6% curry powder

Proportions are derived from dietary survey records.

### 248 Pilau, egg and potato, brown rice

15.6% boiled egg
51.3% boiled brown rice
12.5% onions fried in corn oil
15.6% boiled potatoes

3.1% green chillies
1.3% salt
0.6% curry powder

Proportions are derived from dietary survey records.

### 249 Pilau, mushroom

200g raw white rice
53ml water absorbed on soaking
25g butter ghee
½tsp cumin seeds
90g sliced onion

90g sliced onion
100g sliced mushrooms
500ml boiling water
1tsp salt

Soak the rice and drain. Heat the ghee and add cumin seeds. Add the onion and fry until soft. Add mushrooms and fry for 2 minutes then remove and keep to one side. Add rice and water. Bring to the boil, cover and cook for 20 minutes. Stir in mushrooms and onions continue to cook for a further 10 minutes or until all the water has been absorbed.

Weight loss: 23%

### 250 Pilau, plain

200g raw white rice
45ml water absorbed on soaking
25g butter ghee
1tsp cumin seeds

90g sliced onion
500ml boiling water
1tsp salt
½tsp turmeric

Soak the rice. Heat the ghee and add cumin seeds. Add the onion and fry for 5 minutes. Reduce heat, add remaining ingredients, cover and cook until all the water has been absorbed and the rice is soft.

Weight loss: 17%

### 251 Pilau, vegetable

200g raw white rice
60ml water absorbed on soaking
25g butter ghee
1tsp cumin seeds
90g sliced onion

90g sliced onion
100g frozen mixed vegetables
500ml boiling water
1tsp salt
½tsp turmeric

Method as pilau, plain (No. 250).

Weight loss: 22%

### 252 Pizza, cheese and tomato

370g pizza dough
200g tomatoes
150g cheese

40g black olives
20g vegetable oil

Press dough into a pizza tin and leave in a warm place. Arrange sliced or pulped tomatoes on top followed by cheese and olives. Brush with oil. Bake at 230°C/mark 8 for 30 minutes.

Weight loss: 14%

### 253 Pizza, cheese and tomato, wholemeal

370g wholemeal pizza dough    40g black olives
200g tomatoes    20g vegetable oil
150g cheese

Method as pizza, cheese and tomato (No. 252).

Weight loss: 14%


### 255 Pizza, tomato

200g pizza dough    5g tomato purée
44g olive oil    15g chopped garlic
250g tomatoes    ¼tsp salt

Press dough into a pizza tin and leave in a warm place. Brush with half the oil. Remove skin and seed from the tomatoes and drain add and blend with tomato purée. Spread mixture over the pizza base and sprinkle over garlic. Pour over remaining oil and bake at 180°C/mark 4 for 20 minutes.

Weight loss: 15%


### 256 Pizza, tomato, wholemeal

200g wholemeal pizza dough    5g tomato purée
44g olive oil    15g chopped garlic
250g tomatoes    ¼tsp salt

Method as pizza, tomato (No. 255).

Weight loss: 15%


### 257 Potato cakes, fried in lard

450g mashed potatoes    ½tsp salt
170g flour    ½tsp pepper
15g butter    lard absorbed on frying (40g)
60ml milk

Mix the potatoes, flour and butter with the milk until consistency is smooth. Add seasoning and shape into rounds approximately 2cms thick. Fry in lard until brown.

Weight loss: 9%


### 258 Potato cakes, fried in vegetable oil

450g mashed potatoes    ½tsp salt
170g flour    ½tsp pepper
15g butter    vegetable oil absorbed on frying (40g)
60ml milk

Method as potato cakes, fried in lard (No. 257).

Weight loss: 9%

259 **Potato, leek and celery bake**

| | |
|---|---|
| 40% potato | 2% skimmed milk powder |
| 23% water | 4% butter |
| 8% cheese | 2% flour |
| 8% leek | 1% salt |
| 8% celery | 1% pepper |
| 3% cornflour | 1% mustard |

Proportions are derived from dietary survey records.


260 **Potatoes, duchesse**

500g boiled potatoes
25g butter
1 egg

Sieve or mash potatoes and then beat in butter and egg. Cool. Pipe into shape. Cook at 200°C/mark 6 for approximately 25 minutes until browned and set.

Weight loss: 13%


261 **Potatoes with eggs**

| | |
|---|---|
| 90g chopped onion | 4 beaten eggs |
| 44g corn oil | 5g salt |
| 190g fried diced potatoes | 2g pepper |

Fry onion in the oil, add potatoes, salt, pepper and beaten eggs. Stir until eggs and potato are well blended and the egg is cooked.

Weight loss: 16%


262 **Quorn korma**

| | |
|---|---|
| water | vegetable oil |
| rice | tomato purée |
| Quorn | modified starch |
| onion | seasoning |
| creamed coconut | dessicated coconut |
| whipping cream | salt |
| lemon juice | ground almonds |
| flaked almonds | |

Proportions obtained from a manufacturer

Weight loss: 12% on re-heating


263 **Ratatouille**

| | |
|---|---|
| 150g chopped onion | 230g sliced courgettes |
| 5g garlic | 135g sliced green pepper |
| 65g vegetable oil | 255g skinned and chopped tomatoes |
| 420g sliced aubergine | 5g salt |

Fry onion and garlic in the oil until soft. Add remaining vegetables, stir and cover. Simmer gently for 50 minutes until tender.

Weight loss: 22%

### 265 Re-fried beans

835g boiled red kidney beans
750ml water
90g chopped onion
5g crushed garlic

10g vegetable oil
1tsp chilli powder
100g margarine
1tsp salt

Place beans in a saucepan together with water, onion, garlic, oil and chilli powder. Bring to the boil and then simmer until beans are tender and liquid absorbed. Add the margarine and salt and mash thoroughly.

Weight loss: 43%


### 266 Red pea loaf

885g boiled red kidney beans
15g vegetable oil
2 beaten eggs

5g salt
4g pepper
15g grated cheese

Mash the beans and rub through a sieve. Mix with oil, eggs and seasonings. Pack into a lightly greased dish, sprinkle with cheese and bake at 190°C/mark 5 for 15-20 minutes.

Weight loss: 3%


### 267 Rice and black-eye beans

225g dry black-eye beans
295ml water absorbed on soaking
  and boiling beans
568ml water
5g crushed garlic
20g creamed coconut

½tsp thyme
1tsp salt
70g chopped onion
25g margarine
455g raw white rice
60ml water(adhered to washed rice)

Soak beans overnight. Drain add fresh water, garlic and coconut and cook until soft. Adjust water to specified quantity. Add seasoning, onion and margarine. Wash rice, add to bean mixture, bring to boil and then simmer. When most of the water has evaporated reduce heat, cover tightly and allow to steam cook.

Weight loss: 12%


### 268 Rice and black-eye beans, brown rice

225g dry black-eye beans
295ml water absorbed on soaking
  and boiling beans
568ml water
5g crushed garlic
20g creamed coconut

½tsp thyme
1tsp salt
70g chopped onion
25g margarine
455g raw white rice
60ml water(adhered to washed rice)

Method as rice and black-eye beans (No.267).

Weight loss: 12%

### 269  Rice and pigeon peas

225g dry whole pigeon peas  
320ml water absorbed on soaking  
  and boiling peas  
668ml water  
5g garlic  
20g creamed coconut  

½tsp thyme  
1tsp salt  
70g chopped onion  
25g margarine  
455g raw white rice  
60ml water(adhered to washed rice)  

Method as rice and black-eye beans (No. 267).

Weight loss: 14%


### 270  Rice and pigeon peas, brown rice

225g dry whole pigeon peas  
320ml water absorbed on soaking  
  and boiling peas  
668ml water  
5g crushed garlic  
20g creamed coconut  

½tsp thyme  
1tsp salt  
70g chopped onion  
25g margarine  
455g raw brown rice  
60ml water(adhered to washed rice)  

Method as as rice and black-eye beans (No. 267).

Weight loss: 14%


### 271  Rice and red kidney beans

225g dry red kidney beans  
345ml water absorbed on soaking  
  beans  
610ml water  
20g creamed coconut  

½tsp thyme  
1tsp salt  
25g margarine  
455g raw white rice  
60ml water(adhered to washed rice)  

Soak peas overnight. Adjust water to specified quantity. Add coconut, thyme and salt and cook until soft. Add margarine and rice. Cover and cook.

Weight loss: 11%


### 272  Rice and red kidney beans, brown rice

225g dry red kidney beans  
345ml water absorbed on soaking  
  beans  
610ml water  
20g creamed coconut  

½tsp thyme  
1tsp salt  
25g margarine  
455g raw brown rice  
60ml water(adhered to washed rice)  

Method as rice and red kidney beans (No. 271).

Weight loss: 11%

### 273 Rice and split peas

225g dry split peas
230ml water absorbed on soaking
  and boiling peas
568ml water
5g crushed garlic
20g creamed coconut

½tsp thyme
1tsp salt
70g chopped onion
25g margarine
455g raw white rice
60ml water(adhered to washed rice)

Method as for rice and black-eye beans (No. 267).

Weight loss: 15%

### 274 Rice and split peas, brown rice

225g dry split peas
230ml water absorbed on soaking
  and boiling peas
568ml water
5g crushed garlic
20g creamed coconut

½tsp thyme
1tsp salt
70g chopped onion
25g margarine
455g raw brown rice
60ml water(adhered to washed rice)

Method as for rice and black-eye beans (No. 267).

Weight loss: 15%

### 275 Risotto, vegetable

90g chopped onion
44g vegetable oil
175g raw white rice
15g crushed garlic
600ml water
1tsp salt

60g thinly sliced celery
160g diced red pepper
250g sliced mushrooms
270g canned red kidney beans
15g soya sauce
50g cashew nuts

Fry onion in half the oil, add rice and some of the garlic, cook with stirring for 3 minutes. Add water and salt, bring to the boil, cover and simmer for 30-40 minutes until all the water has been absorbed. Fry celery, pepper and mushrooms in the remaining oil until soft, add rest of garlic. Add rice mixture, kidney beans, soya sauce and nuts. Cook until the beans are heated through.

Weight loss: 31%

### 276 Risotto, vegetable, brown rice

90g chopped onion
44g vegetable oil
175g raw brown rice
15g crushed garlic
600ml water
1tsp salt

60g thinly sliced celery
160g diced red pepper
250g sliced mushrooms
270g canned red kidney beans
15g soya sauce
50g cashew nuts

Method as risotto, vegetable (No. 275).

Weight loss: 31%

### 277 Rissoles, chick pea, fried in sunflower oil

450g canned or boiled chick peas    30g beaten egg
5g chopped fresh parsley    sunflower oil absorbed on frying (60g)

Mash or purée the chick peas. Mix together with the parsley and beaten egg. Shape and fry in oil.

Weight loss: 17%

### 278 Rissoles, chick pea, fried in vegetable oil

450g canned or boiled chick peas    30g beaten egg
5g chopped fresh parsley    vegetable oil absorbed on frying (60g)

Mash or purée the chick peas. Mix together with the parsley and beaten egg. Shape and fry in oil.

Weight loss: 17%

### 279 Rissoles, lentil, fried in sunflower oil

90g chopped onion    ½tsp salt
90g grated carrot    ½tsp pepper
22g sunflower oil    1 egg, beaten
225g dry whole lentils    140g wholemeal breadcrumbs
490ml water    sunflower oil absorbed on frying (55g)
1tsp ground coriander

Fry the onion and carrot in oil until soft. Add lentils, water, coriander, pepper and salt and pepper and simmer until lentils have cooked. Mix in about a third of the breadcrumbs, add the egg and shape into rissoles. Coat with remaining breadcrumbs. Shallow fry until golden brown.

Weight loss: 25%

### 280 Rissoles, lentil, fried in vegetable oil

90g chopped onion    ½tsp salt
90g grated carrot    ½tsp pepper
22g vegetable oil    1 beaten egg
225g dry whole lentils    140g wholemeal breadcrumbs
490ml water    vegetable oil absorbed on frying (55g)
1tsp ground coriander

Method as rissoles, lentil fried in sunflower oil (No. 279).

Weight loss: 25%

### 281 Rissoles, rice, fried in sunflower oil

43.6% boiled brown rice    0.5% curry powder
32.7% wholemeal breadcrumbs    0.3% salt
9.8% onion    sunflower oil absorbed on frying (6.5%)
6.5% egg

Proportions are derived from dietary survey records.

## 282 Rissoles, rice, fried in vegetable oil

43.6% boiled brown rice
32.7% wholemeal breadcrumbs
9.8% onion
6.5% egg

0.5% curry powder
0.3% salt
vegetable oil absorbed on frying (6.5%)

Proportions are derived from dietary survey records.


## 283 Rissoles, vegetable, fried in sunflower oil

58.7% boiled potato
15% boiled turnip
15% boiled carrot
1.9% boiled peas

1.9% flour
1.9% margarine
sunflower oil absorbed on frying (5.6%)

Proportions are derived from dietary survey records.


## 284 Rissoles, vegetable, fried in vegetable oil

58.7% boiled potato
15% boiled turnip
15% boiled carrot
1.9% boiled peas

1.9% flour
1.9% margarine
vegetable oil absorbed on frying (5.6%)

Proportions are derived from dietary survey records.


## 285 Roulade, spinach

420g boiled spinach
240g eggs(approx 5)
45g butter
3g salt

1g pepper
30g flour
284ml semi-skimmed milk
150g cheese

Quantities are derived from dietary survey records.

Weight loss: 24%


## 286 Salad, bean, retail

canned green beans
canned black-eye beans
canned red kidney beans
French dressing
canned sweetcorn

canned chick peas
canned haricot beans
red pepper
onion

Proportions obtained from a manufacturer.


## 287 Salad, beetroot

500g cooked beetroot
90g chopped or sliced onion
60ml French dressing

Slice the beetroot, mix with chopped or sliced onion. Cover with French dressing.

## 288 Salad, carrot and nut, with French dressing, retail

carrot
French dressing
peanuts

sultanas
sugar

Proportions obtained from a manufacturer


## 289 Salad, carrot and nut, with mayonnaise, retail

carrot
mayonnaise
sultanas

low fat natural yogurt
peanuts

Proportions obtained from a manufacturer


## 290 Salad, Florida, retail

white cabbage
mayonnaise
canned mandarins
canned pineapple

apple
celery
glacé cherries

Proportions obtained from a manufacturer.


## 291 Salad, Greek

300g tomatoes
160g sliced green pepper
230g sliced cucumber
140g diced feta cheese

50g black olives, stones removed
90g olive oil
30ml lemon juice
1g salt

Cut tomatoes into quarters. Mix with sliced pepper and cucumber, feta cheese and olives. Pour over the oil, lemon juice and add salt. Toss.


## 292 Salad, green

150g shredded lettuce
230g sliced cucumber

160g sliced green pepper
30g sliced celery

Toss all ingredients together.


## 293 Salad, pasta

480g boiled pasta
140g green pepper
60g spring onion
150g carrot

60g peas
90g mayonnaise
5g fresh parsley

Quantities are derived from dietary survey records.

### 294 Salad, pasta, wholemeal

| | |
|---|---|
| 480g boiled wholemeal pasta | 60g peas |
| 140g green pepper | 90g mayonnaise |
| 60g spring onion | 5g fresh parsley |
| 150g carrot | |

Quantities are derived from dietary survey records.


### 295 Salad, potato, with French dressing

| | |
|---|---|
| 630g boiled potatoes | 50g chopped onions |
| 125ml French dressing | 10g chopped fresh parsley |

Dice potatoes, place in bowl with onion and parsley. Pour over the dressing and toss well.


### 296 Salad, potato, with mayonnaise

| | |
|---|---|
| 630g boiled potatoes | 50g chopped onion |
| 260ml mayonnaise | 10g chopped fresh parsley |

As salad, potato with French dressing (No. 295).


### 297 Salad, potato with mayonnaise, retail

potato
mayonnaise
onion

Proportions obtained from a manufacturer.


### 298 Salad, potato, with reduced calorie dressing, retail

potato
reduced calorie dressing
onion

Proportions obtained from a manufacturer.


### 299 Salad, rice

| | |
|---|---|
| 720g boiled white rice | 60g raisins |
| 240g spring onion | 40g soya sauce |
| 90g sweetcorn | 60g vegetable oil |
| 60g cashew nuts | 80g green pepper |

Quantities are derived from dietary survey records.


### 300 Salad, rice, brown

| | |
|---|---|
| 720g boiled brown rice | 60g raisins |
| 240g spring onion | 40g soya sauce |
| 90g sweetcorn | 60g vegetable oil |
| 60g cashew nuts | 80g green pepper |

Quantities are derived from dietary survey records.

### 301 Salad, tomato and onion

250g sliced tomatoes
90g sliced onion

30ml French dressing

Arrange the tomatoes and onions in a dish. Spoon over the French dressing.


### 302 Salad, vegetable, canned

60g canned potato
14g canned garden peas
40g canned carrots

8g canned sweetcorn
12g canned runner beans
60g salad cream

Quantities are derived from dissection of shop-bought samples.


### 303 Salad, Waldorf

75g chopped celery
400g diced apples
40g walnut halves

115g mayonnaise
20ml lemon juice

Mix celery, apple and walnuts. Add mayonnaise and lemon juice and toss.


### 304 Salad, Waldorf, retail

mayonnaise
celery
apple
cabbage

walnuts
raisins
lemon juice

Proportions obtained from a manufacturer.


### 306 Sauce, curry, onion, with butter

300g chopped onions
120g butter
10g chopped green chillies

8g curry powder
100ml water

Fry onions in the butter until brown. Add remaining ingredients and simmer.

Weight loss: 25%


### 307 Sauce, curry, onion, with vegetable oil

300g chopped onions
120g vegetable oil
10g chopped green chillies

8g curry powder
100ml water

Method as sauce, curry, onion, with butter (No. 306).

Weight loss: 25%

### 308 Sauce, curry, sweet

10g vegetable oil
50g chopped onion
10g flour
5g curry powder
5g tomato purée

375ml water
25g chopped apple
5g desiccated coconut
10g sultanas

Heat the oil and fry the onion. Mix in flour and curry powder, cook gently. Mix in the tomato purée and cool. Gradually mix in the water to form a smooth sauce. Add remaining ingredients and simmer for 30 minutes.

Weight loss: 50%

### 309 Sauce, curry, tomato and onion

100g finely chopped onions
5g crushed garlic
10g chopped root ginger
50g vegetable oil
½tsp ground coriander

½tsp red chilli powder
½tsp turmeric
226g canned tomatoes
40ml water
½tsp salt

Fry onions, garlic and ginger in oil. Add spices and simmer for 2 minutes, add water, tomatoes and salt and cook gently for 20 minutes.

Weight loss: 39%

### 310 Sauce, tomato base

440g canned tomatoes
22g olive oil
70g finely chopped onion
2g crushed garlic

20g tomato purée
½tsp sugar
5g fresh basil

Chop tomatoes. Heat oil, add onions and garlic and cook until soft. Add remaining ingredients, cover and simmer gently for 30 minutes.

Weight loss: 20%

### 311 Sauce, tomato and mushroom

400g canned tomatoes
80g onions fried in corn oil

30g mushrooms fried in corn oil
1tsp mixed herbs

Quantities are derived from a recipe review.

Weight loss: 20%

### 312 Shepherd's pie, vegetable

90g chopped onion
30g chopped celery
50g grated carrot
110g sliced mushrooms
50g vegetable oil

530g whole boiled lentils
30g tomato purée
1tsp mixed herbs
½tsp salt
680g mashed potato

Fry the vegetable in the oil until tender. Add lentils, tomato purée, herbs and salt and mix well. Pour into dish, top with mashed potato. Bake at 200°C/mark 6 for 30-40 minutes.

Weight loss: 11%

### 313 **Shepherd's pie, vegetable, retail**

boiled potatoes
tomatoes
boiled onion
water
boiled lentils
boiled carrots
boiled courgettes

single cream
boiled pearl barley
butter
soya oil
tomato purée
corn starch
salt

Proportions obtained from a manufacturer.

Weight loss: 11% on re-heating


### 315 **Sweet potato and onion layer**

900g sweet potatoes
115g thinly sliced onions
½tsp salt
½tsp pepper

56g flour
350ml milk
30g grated cheese

Parboil potatoes, peel and slice thinly. Place potatoes and onions in layers in a lightly oiled dish, sprinkling each layer with salt and pepper. Make a sauce with the milk and flour, add seasoning and pour over the vegetables. Sprinkle with cheese and bake at 180°C/mark 4 for 30 minutes.

Weight loss: 24%


### 316 **Tabouleh**

115g bulgur wheat
265g water absorbed on soaking
80g chopped spring onions
20g chopped fresh parsley

10g chopped fresh mint
22g vegetable oil
20ml lemon juice
¼tsp salt

Soak the wheat for about an hour, drain thoroughly. Mix with remaining ingredients.


### 317 **Tagliatelle, with vegetables, retail**

water
boiled tagliatelle
tomatoes
boiled onions
boiled courgettes
cream
milk

boiled aubergines
soya oil
modified starch
garlic purée
sugar
starch

Proportions obtained from a manufacturer.

Weight loss: 9% on re-heating

### 318 Tempeh burgers, fried in vegetable oil

90g chopped onion  
11g vegetable oil  
200g tempeh  
500g boiled brown rice  

½tsp salt  
1 egg  
21g flour  
vegetable oil absorbed on frying (32g)  

Fry the onion in a little oil. Finely mash or purée tempeh, rice, salt and beaten egg. Add onion. Form into burger shapes. Coat with flour and shallow fry until brown.

Weight loss: 9%

### 319 Tofu burger, baked

350g tofu  
75g rolled oats  
45g grated carrot  
90g chopped onion  

1tsp mixed herbs  
20g soya sauce  
25g wheatgerm  

Mix all ingredients except wheatgerm into a firm blend and let stand. Shape into 8 burgers and coat with wheatgerm. Bake at 180°C/mark 4 for 15-20 minutes.

Weight loss: 6%

### 320 Tofu spread

250g tofu  
70g onion  
60g celery  
20g soya sauce  

1tsp salt  
1tsp paprika  
5g fresh parsley  
130g mayonnaise  

Crumble tofu into a liquidizer. Add all the ingredients except the mayonnaise. Blend until smooth and stir in mayonnaise.

### 321 Tomatoes, stuffed with rice

325g tomato cases, pulp removed  
9g sugar  
70g olive oil  
45g chopped onion  
95g raw white rice  
65g tomato pulp  
70ml water  

30g currants  
1g salt  
5g fresh chopped parsley  
120g tomato pulp  
10g olive oil  
20g breadcrumbs  

Sprinkle cases with the sugar. Heat oil and fry the onion. Add rice and cook for a few minutes. Add half the tomato pulp and all the water, currants, seasoning and parsley. Cover and cook for 10 minutes. Fill tomatoes with the mixture leaving room for the rice to swell. Force remaining tomato pulp through a sieve and into the tomatoes. Cover with tomato lids. Pour a little oil over each tomato and sprinkle with the breadcrumbs. Bake at 180°C/mark 5 for 1½ hours.

Weight loss: 12% for filling

Weight loss: 22% for baking stuffed tomatoes

### 322 **Tomatoes, stuffed with vegetables**

510g tomatoes
25g vegetable oil
60g chopped onion
70g sliced mushrooms

140g canned sweetcorn
½tsp salt
½tsp mixed herbs
14g tomato purée

Cut tops from tomatoes and scoop out pulp. Fry onion in oil, add tomato pulp, mushrooms and continue. Add remaining ingredients and simmer until reduced. Fill tomatoes with mixture, replace lids and bake at 200°C/mark 6 for 20-30 minutes.

Weight loss: 43%

### 341 **Vegetable bake**

210g carrots
120g courgettes
120g onions
210g potatoes
45g margarine

30g flour
426ml milk
60g Leicester cheese
½tsp mustard powder
45g white breadcrumbs

Quantities are derived from dietary survey records.

Weight loss: 15%

### 347 **Vine leaves, stuffed with rice**

450g chopped onions
170g olive oil
140g raw white rice
112g currants
10g chopped mint

370ml water
¼tsp pepper
15g salt
20ml lemon juice
240g preserved vine leaves (approx 20)

Fry the onions in half of the oil, add rice and cook for 5 minutes. Add currants, mint, seasonings, and one cup of water. Simmer for 5 minutes and cool. Heap into teaspoons and roll into leaves. Place leaves in a pan and sprinkle with lemon juice, remaining oil and water. Cover and simmer over a low heat for an hour.

Weight loss: 36%

# ALTERNATIVE NAMES

The alternative names included in this table are those that were most frequently encountered during recipe collection and are included to help in identifying foods. It is important to recognise that many more names will exist for the dishes below and that some names may be used for more than one food.

Foods are listed below in the same order as in the main tables.

| | Food name | Alternative name |
|---|---|---|
| 9 | **Bhaji, aubergine and potato** | Baingan aloo bhaji<br>Bateta ringna noo shak<br>Brinjal and potato bhaji |
| 11 | **Bhaji, cabbage** | Gobi bhaji<br>Kobi noo shak |
| 12<br>13 | **Bhaji, cabbage and pea** | Gobi matar bhaji<br>Matar kobi noo shak |
| 14<br>15 | **Bhaji, cabbage and potato** | Gobi aloo bhaji<br>Kobi bateta noo shak |
| 16 | **Bhaji, cabbage and spinach** | Gobi palak bhaji<br>Kobi palak noo shak |
| 17<br>18 | **Bhaji, carrot, potato and pea** | Gajar aloo matar bhaji<br>Gajar bateta matar noo shak |

| | Food name | Alternative name |
|---|---|---|
| 20 | **Bhaji, cauliflower and potato** | Aloo gobi<br>Aloo phool kobi<br>Bateta phool kobi noo shak<br>Rhulkobi aloo bhaji |
| 28 | **Bhaji, mustard leaves** | Saag<br>Saag bhaji<br>Sanson-ka sag |
| 29 | **Bhaji, mustard leaves and spinach** | Saag<br>Saag bhaji<br>Sanson palak bhaji |
| 30<br>31<br>32 | **Bhaji, okra** | Bhinda bhaji<br>Bhinda noo shak<br>Bhindee bhajee<br>Bhindi bhaji |
| 33 | **Bhaji, pea** | Matar bhaji<br>Matar noo shak |
| 34<br>35 | **Bhaji, potato** | Aloo bhaji<br>Bateta noo shak |
| 36 | **Bhaji, potato and fenugreek leaves** | Aloo methi<br>Bateta methi noo shak<br>Methi bhaji aloo |
| 38 | **Bhaji, potato and onion** | Aloo piyaz bhaji<br>Bateta doongali noo shak |

| | Food name | Alternative name |
|---|---|---|
| 39 | **Bhaji, potato, onion and mushroom** | Mashroom bateta noo shak<br>Mushroom aloo bhaji |
| 41 | **Bhaji, spinach** | Palak ni bhaji<br>Palak sag<br>Sag<br>Saag<br>Saag bhaji |
| 42 | **Bhaji, spinach and potato** | Palak aloo bhaji<br>Palak bateta noo shak |
| 76 | **Coco fritters** | Eddoe fritters<br>Tannia fritters |
| 81 | **Corn fritters** | Sweetcorn fritters |
| 88 | **Curry, almond** | Badam<br>Badaam |
| 93 | **Curry, black gram whole and red kidney bean** | Rajmaan and maan<br>Rajmah and aalad |
| 94<br>95 | **Curry, black-eye bean** | Chora<br>Lobia curry |
| 98 | **Curry, cauliflower and potato** | Phool kobi bateta noo shak |
| 100<br>101 | **Curry, chick pea dahl and spinach** | Chana dahl and palak |

| | Food name | Alternative name |
|---|---|---|
| 102 | **Curry, chick pea, whole** | Chana<br>Chola<br>Kabuli chana |
| 104 | **Curry, chick pea, whole,<br> and potato** | Aloo chana<br>Chana aloo<br>Chana bateta |
| 105<br>106<br>107<br>108 | **Curry, chick pea, whole<br> and tomato** | Chana |
| 110 | **Curry, dudhi kofta** | Dudhi na kofta |
| 116 | **Curry, green bean** | Fansi noo shak |
| 118<br>119<br>120 | **Curry, lentil, red/masoor dahl** | Masaran di dhoti dahl<br>Masur dahl |
| 127<br>128 | **Curry, lentil whole/masoor** | Akha masoor<br>Sabut masoor |
| 129<br>130 | **Curry, mung bean dahl** | Moong dahl<br>Moong dhoti dahl<br>Moong ni dahl |
| 133<br>134 | **Curry, mung bean, whole** | Akha moong<br>Moong sabat dahl |

| | Food name | Alternative name |
|---|---|---|
| 136 | **Curry, okra** | Bhinda noo shak<br>Masala bhindi<br>Masala okra |
| 138<br>139 | **Curry, pigeon pea dahl** | Toor dahl<br>Toovar dahl<br>Tuvar dahl |
| 140 | **Curry, pigeon pea dahl<br>and tomato** | Toor dahl<br>Toovar dahl<br>Tuvar dahl |
| 141 | **Curry, pigeon pea dahl with<br>tomatoes and peanuts** | Toor dahl<br>Toovar dahl<br>Tuvar dahl |
| 147 | **Curry, red kidney and mung<br>bean whole** | Rajmah ana moong<br>Rajmea and moong |
| 148 | **Curry, spinach** | Masala spinach<br>Palak ni bhaji<br>Saag |
| 149 | **Curry, spinach and potato** | Palak aloo<br>Palak bateta |
| 160 | **Dosa, plain** | Dona<br>Pancake<br>Dosmama salea wala |
| 181 | **Khadhi** | Kadhi |

| | Food name | Alternative name |
|---|---|---|
| 183 | **Khichadi** | Khichiri |
| 184 | | |
| 194 | **Lentil cutlets** | Masoor dahl cutlets |
| 222 | **Okra with tomatoes and onion, Greek** | Bamyes |
| 224 | **Pakora/bhajia, aubergine** | Pakorea<br>Ringna na bhajiya |
| 225 | **Pakora/bhajia, cauliflower** | Pakorea<br>Phool kobi na bhajiya |
| 226 | **Pakora/bhajia, onion** | Doongali na bhajiya<br>Pakorea |
| 227 | **Pakora/bhajia, onion, retail** | Doongali na bhajiya<br>Pakorea |
| 228 | **Pakora/bhajia, potato** | Bateta na bhajiya<br>Pakorea |
| 229 | **Pakora/bhajia, potato and cauliflower** | Bateta, phool kobi na bhajiya<br>Pakorea |
| 230 | **Pakora/bhajia, potato, carrot and pea** | Bateta, gajar, matar na bhajiya<br>Pakorea |

| | Food name | Alternative name |
|---|---|---|
| 231 | **Pakora/bhajia, spinach** | Pakorea<br>Palak na bhajiya |
| 232 | **Pakora/bhajia, vegetable, retail** | Pakorea |
| 247<br>248 | **Pilau, egg and potato** | Pulao |
| 249 | **Pilau, mushroom** | Pulao |
| 250 | **Pilau, plain** | Pulao |
| 251 | **Pilau, vegetable** | Pulao |
| 265 | **Re-fried beans** | Frijoles refritos |
| 267<br>268 | **Rice and black-eye beans** | Charwal matar<br>Rice and peas |
| 269<br>270 | **Rice and pigeon peas** | Charwal matar<br>Rice and peas |
| 271<br>272 | **Rice and red kidney beans** | Charwal rajam<br>Rice and peas |
| 273<br>274 | **Rice and split peas** | Charwal matar<br>Rice and peas |

| | Food name | Alternative name |
|---|---|---|
| 306 | **Sauce, curry, onion** | Masala doongree |
| 307 | | Masala onion |
| | | Masala payaz |
| 321 | **Tomatoes, stuffed with rice** | Tomates yemistes me rizi |
| 347 | **Vine leaves, stuffed with rice** | Dolmathakia me rizi |

# INGREDIENT CODES AND ALTERNATIVE NAMES

The full range of ingredients occuring in the recipes are listed below in alphabetical order.

The code numbers on the left hand side have been taken from the earlier food group supplements and Fifth Edition of 'The Composition of Foods'. Only one code has been given per food, representing the most recent occurrence of a food in the publication series. The codes will therefore, direct users to the most relevant data sets used in the recipe calculations. For a complete list of all food codes used in the Fifth Edition and the four supplements, users are referred to the Food Number Index (RSC 1992). Codes from the supplement publications include the 2-digit prefix appropriate to that supplement. Where no prefix is included the data are taken from the Fifth Edition.

The prefixes for each supplement are:

| | |
|---|---|
| Cereals and Cereal Products | 11 |
| Milk Products and Eggs | 12 |
| Vegetables, Herbs and Spices | 13 |
| Fruit and Nuts | 14 |

The alternative names listed in the right hand column are those that were most frequently encountered during data collection. They have been included here to help identify foods used as ingredients and assist users in their interpretation of the recipes.

| Code number | Ingredient name | Alternative names |
|---|---|---|
| 14-801 | **Almonds** | Badaam |
| | | Badam |
| 14-002 | **Apples, cooking, raw** | Saffarjan |
| 14-014 | **Apples, eating, raw** | Saib |
| | | Tarel |

| Code number | Ingredient name | Alternative names |
| --- | --- | --- |
| 13-803 | **Asafoetida** | Heeng<br>Hing<br>Kari sing |
| 739 | **Aubergine, raw** | Baingan<br>Brinjal<br>Eggplant<br>Reengan<br>Ringana |
| 14-037 | **Avocado** | Alligator pear<br>Butter pear<br>Zaboca |
| 1174 | **Baking powder** | Pakane ka soda |
| 13-804 | **Basil, fresh** | Sweet basil<br>Tea bush<br>Tulsi |
| 693 | **Beans, aduki, boiled** | Adzuki beans |
| 700<br>701 | **Beans, black-eye, dry**<br>**Beans, black-eye, boiled** | Blackeye peas<br>Cowpeas<br>Chori<br>Lobia<br>Rawan<br>Rongi |
| 698 | **Beans, black gram whole, dry** | Alad<br>Black matape<br>Urad<br>Urd sabat |

| Code number | Ingredient name | Alternative names |
| --- | --- | --- |
| 13-060 | **Beans, black gram dahl, dry** | Duhli urad<br>Urd dali hoi |
| 703 | **Beans, butterbeans, canned** | |
| 707<br>708<br>13-085 | **Beans, green, raw**<br>**Beans, green, frozen, boiled**<br>**Beans, green, canned** | Fansi<br>Hari saim<br>Kataud<br>Shim |
| 13-087 | **Beans, haricot, boiled** | Kataud<br>Saim ki pali |
| 714<br>715 | **Beans, mung whole, dry**<br>**Beans, mung, whole, boiled** | Green gram<br>Golden gram<br>Moong beans |
| 13-098<br>13-099 | **Beans, mung, dahl, dry**<br>**Beans, mung, dahl, boiled** | Moongi dhoti |
| 13-102<br>13-103 | **Beans, pigeon peas, whole, dry**<br>**Beans, pigeon peas,boiled** | Arhar<br>Congo peas<br>Gunga peas<br>Red gram<br>Tuvar |
| 13-104<br>13-105 | **Beans, pigeon peas dahl, dry**<br>**Beans, pigeon, boiled** | Masoor ni daar<br>Tuvar dahl |
| 716<br>717<br>718 | **Beans, red kidney, dry**<br>**Beans, red kidney, boiled**<br>**Beans, red kidney, canned** | |

| Code number | Ingredient name | Alternative names |
|---|---|---|
| 719 | **Beans, runner, raw** | |
| 720 | **Beans, runner, boiled** | |
| 722 | **Beans, soya, boiled** | |
| 742 | **Beetroot, boiled** | Chakander |
| 11-069 | **Breadcrumbs, packet** | |
| 744 | **Broccoli, green, raw** | Calabrese |
| 745 | **Broccoli, green, boiled** | |
| 11-007 | **Bulgur wheat** | |
| 306 | **Butter** | Maakhan |
| 749 | **Cabbage, average, raw** | Badla coppi |
| 750 | **Cabbage, average, boiled** | Band gobi |
| | | Bund gobhi |
| | | Kobi |
| 13-190 | **Cabbage, red, raw** | |
| 753 | **Cabbage, white, raw** | |
| 13-809 | **Cardamom** | |

| Code number | Ingredient name | Alternative names |
|---|---|---|
| 754 | **Carrots, raw** | Gajar |
| 755 | **Carrots boiled** | Gajjar |
| 758 | **Carrots, canned** | |
| | | |
| 14-811 | **Cashew nuts, plain** | Kaaju |
| | | Kaju |
| | | |
| 759 | **Cauliflower, raw** | Pangoli |
| 760 | **Cauliflower, boiled** | Phool gobhi |
| | | Pul gobi |
| | | |
| 761 | **Celery, raw** | |
| 762 | **Celery boiled** | |
| | | |
| 228 | **Cheese, Cheddar** | Paneer |
| 232 | **Cheese, cottage, plain** | |
| 238 | **Cheese, feta** | |
| 12-167 | **Cheese, Leicester** | |
| 12-170 | **Cheese, Mozzarella** | |
| 247 | **Cheese, Parmesan** | |
| 12-176 | **Cheese, Ricotta** | |
| | | |
| 14-068 | **Cherries, glacé** | |
| | | |
| 13-073 | **Chick pea flour** | Besan flour |
| | | Chanoa nau laut |
| | | |
| 704 | **Chick peas whole, dry** | Channa |
| 705 | **Chick peas, boiled** | Common gram |
| 706 | **Chick peas, canned** | Garbanzo |
| 13-076 | **Chick peas, dahl, dry** | Yellow gram |

| Code number | Ingredient name | Alternative names |
|---|---|---|
| 838 | **Chilli powder** | |
| 13-814 | **Chives, fresh** | Piaz di kism |
| 13-227 | **Cho cho, raw** | Chayote<br>Chow chow<br>Choko<br>Christophene<br>Mitliton<br>Vegetable pear |
| 839 | **Cinnamon** | Dalchini<br>Tej |
| 13-816 | **Cloves** | |
| 13-376 | **Coco, raw** | Cocoyam<br>Dasheen<br>Eddoe<br>Taro |
| 14-818 | **Coconut, desiccated** | Copru<br>Khopra |
| 14-817 | **Coconut, creamed block** | Cocoanut<br>Nareal<br>Naryal<br>Pharaoh's nut |
| 13-817 | **Coriander leaves, fresh** | Chinese parsley<br>Dhana<br>Dhania |

| Code number | Ingredient name | Alternative names |
| --- | --- | --- |
| 13-819 | **Coriander ground** | Dhania<br>Dhanyia<br>Lila dahna |
| 322 | **Corn oil** | |
| 4 | **Cornflour** | Makaika ata<br>Makai nau laut<br>Nishasta |
| 11-011 | **Cornmeal, sifted** | |
| 764<br>765 | **Courgette, raw**<br>**Courgette, boiled** | Zucchini |
| 212 | **Cream, single** | |
| 767 | **Cucumber** | Kaakdee<br>Kakdi<br>Kheera<br>Khira<br>Tar |
| 13-820 | **Cumin ground** | Jeera<br>Jeeru<br>Sufaid zeera |
| 14-074 | **Currants** | Kaali drakhs |

| Code number | Ingredient name | Alternative names |
|---|---|---|
| 13-821 | **Curry leaves, fresh** | Curry pak<br>Karipatta<br>Nim leaves |
| 840 | **Curry powder** | |
| 13-824 | **Dill, fresh** | |
| 290<br>293 | **Egg, raw**<br>**Egg boiled** | Anda<br>Indaa |
| 13-243 | **Fenugreek leaves** | Bird's foot<br>Methi<br>Methi ni baajee<br>Pisali methai |
| 13-828 | **Fenugreek seeds** | Methi |
| 14 | **Flour, white, plain** | Laut<br>Maida |
| 15 | **Flour, self-raising** | Laut<br>Maida |
| 16 | **Flour, wholemeal** | Ata<br>Atta |
| 841 | **Garam masala** | |

| Code number | Ingredient name | Alternative names |
|---|---|---|
| 772 | **Garlic, raw** | Lassan<br>Lasin<br>Lehsan<br>Roshune |
| 335 | **Ghee, butter** | Clarified butter |
| 337 | **Ghee, vegetable** | |
| 13-832 | **Ginger powder** | Sundh |
| 13-247 | **Ginger root** | Adoo<br>Adrak<br>Adu |
| 13-249 | **Gourd, bottle, raw** | Calabash gourd<br>Chapni kaddu<br>Doodhi<br>Dudhi<br>Lawki<br>White gourd |
| 774 | **Gourd, karela, raw** | Bitter gourd<br>Balsam apple |
| 13-256 | **Gourd, tinda, raw** | Gentleman's toes<br>Round gourd<br>Tindla<br>Tindo<br>Tindola |
| 13-279 | **Green banana, raw** | Matoki |

| Code number | Ingredient name | Alternative names |
| --- | --- | --- |
| 1001 | **Honey** | |
| 1003 | **Jaggery** | |
| 318 | **Lard** | |
| 11-051<br>11-052 | **Lasagne, raw**<br>**Lasagne, boiled** | |
| 776 | **Leeks, boiled** | Dholi leek<br>Payaz hari |
| 14-277 | **Lemon juice** | |
| 712<br>713 | **Lentils, red/masoor dahl, dry**<br>**Lentils, red/masoor dahl, boiled** | Masran di dahl<br>Masur dahl |
| 710<br>711 | **Lentils, whole/masoor, dry**<br>**Lentils, whole/masoor, boiled** | Continental lentils<br>Masar sabat harea<br>Masur |
| 777 | **Lettuce, average** | |
| 14-279 | **Lime juice** | |
| 26 | **Macaroni, boiled** | |

| Code number | Ingredient name | Alternative names |
|---|---|---|
| 309 | **Margarine** | |
| 1179 | **Marmite** | |
| 1159 | **Mayonnaise, retail** | |
| 14-826 | **Melon seeds** | |
| 189 | **Milk, whole** | Dudh |
| | | Doodh |
| 185 | **Milk semi-skimmed** | |
| 181 | **Milk, skimmed** | |
| 202 | **Milk, evaporated, whole** | |
| 200 | **Milk, skimmed milk powder** | Sukha doodh |
| 842 | **Mint, fresh** | Fodina |
| | | Padena |
| 14-827 | **Mixed nuts** | |
| 783 | **Mushrcoms, raw** | Khumb |
| 784 | **Mushrooms, boiled** | Muschrooms |
| 787 | **Mushrooms, fried in corn oil** | |
| 13-298 | **Mustard leaves, raw** | Gandlan |
| | | Saag |
| | | Saroon da sag |
| | | Sarson |

| Code number | Ingredient name | Alternative names |
|---|---|---|
| 13-839 | **Mustard seeds** | Rai<br>Rye<br>Saroon |
| 844 | **Nutmeg** | Jaiphal<br>Jaifur |
| 6 | **Oatmeal, porridge** | |
| 789 | **Okra, raw** | Bhendi<br>Bhinda<br>Bhindi<br>Gumbo<br>Lady's fingers |
| 14-173 | **Olives, black** | Aular<br>Zaitoon |
| 324 | **Olive oil** | Zaitun ka tel |
| 792<br>793 | **Onion, raw**<br>**Onion, boiled** | Doongli<br>Dungli<br>Gandea<br>Kanda<br>Payaz<br>Piyaz |
| 795 | **Onion, fried in corn oil** | |
| 14-281 | **Orange juice, freshly squeezed** | |

| Code number | Ingredient name | Alternative names |
|---|---|---|
| 13-842 | **Oregano** | Jari booti di kism |
| 1182 | **Oxo cubes** | |
| 845 | **Paprika** | Wadi lal mirch |
| 846 | **Parsley, fresh** | Ajmoola de patte |
| 799<br>800 | **Parsnip, raw**<br>**Parsnip, boiled** | Gajar di kism |
| 14-831 | **Peanuts, plain** | Groundnuts<br>Monkey nuts<br>Moong puli<br>Shing |
| 11-003 | **Pearl barley, boiled** | |
| 729<br>733<br>13-132 | **Peas, raw**<br>**Peas, canned**<br>**Peas, frozen, raw** | Badla<br>Mattar<br>Mottor shuti<br>Mutter<br>Vatana<br>Wataanaa |
| 13-141 | **Peas, split, dry** | Matar ki dahl |
| 847 | **Pepper, black** | Kali mirch<br>Kara mari<br>Rai |

| Code number | Ingredient name | Alternative names |
|---|---|---|
| 801 | **Pepper, chilli, green, raw** | Hari mirch |
| | | Lila mirchaa |
| 13-317 | **Pepper, chilli, red, raw** | Lal mirch |
| | | Lal mirchaa |
| 802 | **Pepper, green, raw** | Bell pepper |
| 803 | **Pepper, green, boiled** | Hari mirch |
| | | Mota mircha |
| | | Simla mirch |
| | | Sweet pepper |
| 804 | **Pepper, red, raw** | Bell pepper |
| 805 | **Pepper, red, boiled** | Lal mircha |
| | | Mota mircha |
| | | Simla mirch |
| | | Sweet pepper |
| 14-839 | **Pine nuts** | Chilgoza |
| | | Indian nuts |
| | | Paayn ni badaam |
| | | Pignolias |
| | | Pine kernels |
| 664 | **Potatoes, raw** | Aloo |
| 668 | **Potatoes, boiled, unsalted** | Alu |
| 13-013 | **Potatoes, boiled, salted** | Bateta |
| 663 | **Potatoes, canned** | |
| 685 | **Potatoes, fried, diced** | |
| 811 | **Quorn mycoprotein** | |
| 14-242 | **Raisins** | Kismis |
| | | Munacca |

| Code number | Ingredient name | Alternative names |
|---|---|---|
| 328 | **Rapeseed oil** | |
| 18 | **Rice, brown, raw** | Chawal |
| 19 | **Rice, brown, boiled** | Chavel |
| 22 | **Rice, white, long grain, raw** | |
| 23 | **Rice, white, long grain, boiled** | |
| 11-021 | **Rice flour** | |
| 1161 | **Salad cream** | |
| 1184 | **Salt** | Loon |
| | | Methoo |
| | | Nimak |
| 14-844 | **Sesame seeds** | Benniseeds |
| | | Gingelly |
| | | Til |
| 331 | **Soya oil** | |
| 1171 | **Soya sauce** | |
| 30 | **Spaghetti, white, boiled** | |

| Code number | Ingredient name | Alternative names |
| --- | --- | --- |
| 813 | **Spinach, raw** | Callaloo |
| 814 | **Spinach, boiled** | Palak ni baajee |
| 815 | **Spinach, frozen, boiled** | Palak |
| | | Saag |
| 818 | **Spring onions, raw** | |
| 1011 | **Sugar, white** | Chini |
| | | Khaand |
| 14-263 | **Sultanas** | Kishmish |
| | | Suki laidaraaksh |
| 14-845 | **Sunflower seeds** | |
| 819 | **Swede, raw** | Neep (England) |
| | | Rutabaga |
| | | Yellow turnip |
| 824 | **Sweetcorn kernels, canned** | |
| 821 | **Sweet potato, raw** | Shakaria |
| 822 | **Sweet potato, boiled** | Shakarkundi |
| 14-265 | **Tamarind pulp** | Ambli |
| | | Amli |
| | | Imli |
| | | Indian date |
| 13-118 | **Tempeh** | |

| Code number | Ingredient name | Alternative names |
|---|---|---|
| 827 | **Tomatoes, raw** | Tamater |
| 832 | **Tomatoes, canned** | Tametaa |
| 1172 | **Tomato ketchup** | |
| 826 | **Tomato pureé** | |
| 723 | **Tofu** | |
| 13-861 | **Turmeric powder** | Haldi |
| 833 | **Turnip, raw** | Neep (Scotland)<br>Shalgam |
| 13-392 | **Turnip leaves, boiled** | |
| 782 | **Vegetables, mixed, boiled** | |
| 333 | **Vegetable oil, blended** | Tel |
| 13-393 | **Vine leaves** | |
| 1185 | **Vinegar** | Sirka |

| Code number | Ingredient name | Alternative names |
|---|---|---|
| 14-850 | **Walnuts** | Akrots<br>Akhrot<br>Badaam<br>Madeira nuts |
| 1186 | **Water** | |
| 17 | **Wheatgerm** | |
| 836<br>837 | **Yam, raw**<br>**Yam, boiled** | Ratulu<br>Yarbi |
| 1187<br>1188 | **Yeast** | |
| 260<br>255 | **Yogurt, natural**<br>**Yogurt, natural, low fat** | Dahee<br>Dahi |

# FOOD INDEX

- Foods are indexed by their food number and **not** by page number.

- Cross references in this index (eg Aloo gobi see **Bhaji, cauliflower and potato**) give access to the individual dishes through this index and to alternative names given in the list on pages 208–215